Placer Gold Mining in Arizona

by the U.S. Dept. of Geological Survey

with an introduction by Kerby Jackson

This work contains material that was originally published in 1972 by the U.S. Dept. of Geological Survey.

Introduction

It has been decades since the Department of Interior released it's important publication "Placer Gold Deposits of Arizona". First released in 1972, this important volume has been out of print and has been unavailable to the mining community since those days, with the exception of expensive original collector's copies and poorly produced digital editions.

It has often been said that "*gold is where you find it*", but even beginning prospectors understand that their chances for finding something of value in the earth or in the streams of the Golden West are dramatically increased by going back to those places where gold and other minerals were once mined by our forerunners. Despite this, much of the contemporary information on local mining history that is currently available is mostly a result of mere local folklore and persistent rumors of major strikes, the details and facts of which, have long been distorted. Long gone are the old timers and with them, the days of first hand knowledge of the mines of the area and how they operated. Also long gone are most of their notes, their assay reports, their mine maps and personal scrapbooks, along with most of the surveys and reports that were performed for them by private and government geologists. Even published books such as this one are often retired to the local landfill or backyard burn pile by the descendents of those old timers and disappear at an alarming rate. Despite the fact that we live in the so-called "Information Age" where information is supposedly only the push of a button on a keyboard away, true insight into mining properties remains illusive and hard to come by, even to those of us who seek out this sort of information as if our lives depend upon it. Without this type of information readily available to the average independent miner, there is little hope that our metal mining industry will ever recover.

This important volume and others like it, are being presented in their entirety again, in the hope that the average prospector will no longer stumble through the overgrown hills and the tailing strewn creeks without being well informed enough to have a chance to succeed at his ventures.

Kerby Jackson
Josephine County, Oregon
July 2015

University of Arizona Bulletin

ECONOMIC SERIES No. 18   JANUARY 15, 1922

Arizona Gold Placers

PREFACE

Except for several unsuccessful attempts to use modern dredges, there has been little placer mining activity in the State since Bulletin No. 10 was printed. Nevertheless it has been thought wise to rewrite that bulletin completely, at the same time adding many new data. The present bulletin is, in the main, a compilation of information that was already in print, but scattered and difficult to find, and the aim is to present a concise and fairly complete account off the history, production, location, development, and present status of Arizona's placer deposits in a form readily available for the use of all inquirers, and especially those who think they have developed a process that will enable them to work the deposits profitably.

There are two main obstacles, more or less peculiar to the state of Arizona, that stand in the way of the successful working of these placers: one, lack of water, is a condition common to practically every district in the State; the other, the cementation of the gravel by caliche or lime, is a deterrent in a great number of areas. Many schemes have been devised to overcome these defficulties, but so far none of them has been very successful. These attempts are recounted in the detailed descriptions of the individual placer fields.

HISTORY

Originally the placer gold existed in veins and gold-bearing rocks that were later broken down by the processes of weathering. The disintegrated material was washed away by the torrential rainfall, and was finally collected and concentrated by the same agency in depressions and stream beds. It is in places of this character, scattered widely throughout Arizona, that the gold is now found; and it occurs both in weathered, dry, loose gravels near the surface, and in caliche, or lime-cemented gravels below.

The existence of this gold may have been known previously to Indians and Mexicans, and Raymond[1] states, somewhat indefinitely, that "gold is said to have been discovered on the Gila River in 1858", probably at a point about twenty miles above its junction with the Colorado, but the authentic, recorded history of Arizona placer mining begins with the discoveries of Captain Pauline Weaver. Weaver was a trader and trapper, working from Fort Yuma up and down the Colorado and into the interior, and prospecting as he went. His search was rewarded in 1862 by the discovery of traces of gold in a gulch seven miles east of La Paz, a deposit that later received the name of the La Paz placers.

He did not long have the ground to himself. During a visit to Fort Yuma he displayed the gold he had found, and immediately several Mexicans set out to search for the new placer field. Within less than a mile south of Weaver's camp they panned some rich dirt from which they secured considerable gold. They returned to Yuma for supplies, and the news of the new "diggings" spread so rapidly that within a few months there were 1500[2] people in La Paz and the surrounding territory. Some very rich ground was found, and a number of remarkably large nuggets were uncovered, while there was hardly a gulch for twenty miles around that did not yield more or less gold.

This successful venture encouraged the prospectors to look for new fields, and a rush east began, which resulted in the discovery of the placers on the west slope of the Dome Rock Mountains, the Plomosa placers, and, still further east, the placers of Yavapai county. In May, 1863,[3] a party under Weaver discovered the placers bearing his name. In the same year, a Mexican in the employ of Jack Swilling, during a trip over the mountain to the Weaver camp, discovered the famous Rich Hill placers. Increased prospecting followed these discoveries and resulted in the finding of the Hassayampa and San Domingo placers, and the deposits along Lynx Creek, Granite Creek, and Big Bug Creek; in fact some placer ground was found and worked in practically all the main gulches of the southern and eastern slopes of the Bradshaw Mountains as far east as Camp Creek. It was about this time that Prescott was founded by the prospectors of Granite Creek and the immediate neighborhood.

[1]Production of Gold and Silver in the U. S. by R. W. Raymond, Vol. III, T. A. I. M. E., p. 203, 1874-5.
[2]Mineral Resources of the States and Territories West of the Rocky Mountains, by J. Ross Browne, 1868, pp. 443-482.
[3]Resources of Arizona, by Hamilton, 1888.

Other placers were discovered from time to time. Those of Quijotoa and Indian Oasis have been worked by the Indians for years, in fact the gold taken from them has been exchanged for supplies with the Tucson merchants for as long as records have been kept. In 1874[1] the Greaterville placers were discovered by A. Smith. In the rush that followed all the richer and more available ground was worked by pan and rocker, with water packed in by Mexicans. These deposits have been the basis of many dredging and hydraulicking schemes, but nothing beyond hand sluicing and rocking has ever been done.

The statement just made is true, in fact, as regards practically all the deposits in the State. Other methods have been proposed, but most of the work has been done by hand. It is in the rainy seasons of the year, when water is available for hand work, and the stream beds have been washed and turned over by the torrential rains and flood waters, that most of the placering is done. In this way all the more important placers of the State have been worked over and over again, yet even now a heavy rain will often reveal banks of gold-bearing gravel that have been over-looked.

PRODUCTION

It is impossible to state accurately the total gold and silver production of the placer deposits of Arizona, because most of the gold was obtained at a time when no records were kept, and all estimates have to be based on information secured later from pioneers of the various districts. At best these appraisals cannot be accurate, since the country was very sparsely settled and subject to the conditions prevailing on a frontier, the deposits were widely distributed, and much of the gold was taken out by individuals; however, they give a good idea of the relative importance of the various districts, and some measure of the total production. The most valuable reports are quoted below.

A letter, from Mr. A. McKay, member of the Territorial legislature from La Paz, to Mr. J. Ross Browne,[2] published in 1868, gives information on the earlier production of the La Paz district. Mr. McKay writes:

"Of the yield of these placers, anything like an approximation to the average daily amount of what was taken out per man would only be guess

---

[1]Mines and Mining West of the Rocky Mountains, by R. W. Raymond, 1875, p. 390.
[2]Resources of the States and Territories West of the Rocky Mts., by J. Ross Browne, 1868.

work. Hundreds of dollars per day to the man was common, and now and again a thousand or more a day. Don Juan Ferra took one nugget from his claim that weighed forty-seven ounces and six dollars. Another party found a chispa weighing twenty-seven ounces. Many others found pieces of from one to two ounces up to twenty, and yet it is contended that the greater proportion of the larger nuggets were never shown . . . . . It is the opinion of those most conversant with the first working of these placers that much the greater proportion of the gold taken out was in nuggets weighing from one dollar up to the size mentioned above . . . . . As has been said above, the gold was large and generally clear of foreign substances . . . . . All that was sold or taken here went for $16 to $17 per ounce. Since the year 1864 until the present, there have been at various times many men at work in these placers, numbering in the winter months hundreds, but in the summer months not exceeding seventy-five or one hundred; all seem to do sufficiently well not to be willing to work for the wages of the country, which are and have been for some time from $30 to $65 per month and found. No inconsiderable amount comes in from these placers now weekly, and only a few days ago I saw, myself, a nugget which weighed $40, clear and pure from foreign substance . . . . . "

"Of the total amount of gold taken from these mines, I am as much at a loss to say what it has been as I was to name the average daily wages of the first years, and as I might differ from those who were among the first in these mines, I do not feel justified in setting up an opinion as against them; I shall, therefore, give the substance of the several opinions which I have obtained from those who were the pioneers of these placers. I have failed to find any one of them whose opinion is that less than $1,000,000 were taken from these diggings within the first year, and in all probability as much was taken out in the following years."

On information secured from similar sources Heikes[1] estimates that the placers of Yuma County between 1860 and 1880 produced from $20,000,000 to $42,000,000 in gold.

Browne states that the gravels of Lynx Creek were worked throughout its entire length, about twelve miles, that those of Big Bug Creek were worked for many miles, that they had paid well, and were still yielding good wages. (1868).

---

[1]Dry Placers in Arizona, by V. C. Heikes. U. S. Geological Survey Mineral Resources for 1912, pp. 255-260.

Hamilton, in writing of Rich Hill, states[1] that gold in the Weaver district was found at bed rock, and pieces of pure metal worth several hundred dollars were picked up. It was also common, he states, to find $5000 to $6000 under a single boulder. One acre in this district was said to have yielded $500,000. Hamilton estimates that up to 1883, $1,000,000 had been produced by the placers of the Weaver district. He also placed the production of the placers of Lynx Creek at the same figure, and he believes that it was the richest gold bearing stream in Arizona.

Sparks[2] gives the gold production of Groom Creek as $3,000,000.

In 1876 Raymond[3] reported that the gold from the Greaterville placers was coarse and that nuggets worth from $35 to $50 were brought to Tucson, the average being about $5. The largest nugget ever reported from the camp weighed thirty-seven ounces.[4]

Mr. J. P. Coyne states[5] that by 1881 all the richer stream gravels of Greaterville had been worked over and that by 1886 the placers were considered worked out. They have, however, been worked in a small way by Mexicans ever since. In 1883[6] the yearly production of the district since discovery was estimated at $12,000 and for 1884[7] the total production was $18,000. Mr. Coyne states that the total production of the district probably amounts to $7,000,000 and, according to Hill, "this figure is corroborated by several old time miners who have been in a position to watch the production of the district."

Arizona placers have been worked persistently since discovery and have yielded something every year. Since 1900 the United States Geological Survey has collected and published figures on the placer gold production of the state, and these figures have been brought together in the accompanying table. It will be seen that the total yield of all the placers of the State in the twenty year period 1900 to 1920 is $785,733.

From the information that the writer has been able to gather from all sources it is estimated that the value of the placer gold production of the State of Arizona is at least $50,000,000. This figure is conservative.

[1]Resources of Arizona, by Patrick Hamilton, 1883.
[2]Yavapai, The Land of Opportunity, by G. M. Sparkes, Arizona Bureau of Mines bulletin No. 59, 1917.
[3]Mines and Mining West of the Rocky Mountains, by R. W. Raymond, 1876, p. 342.
[4]Production of Precious Metals in the United States, by H. C. Burchard, 1884, p. 46.
[5]Notes on the Placer Deposits of Greaterville, Arizona, by J. M. Hill. United States Geological Survey Bulletin No. 430, 1909, pp. 11-22.
[6]Production of the Precious Metals in the United States, by H. C. Burchard, 1883, p. 80.
[7]Idem, 1884, p. 46.

A general survey of the dry placer deposits of the State shows extensive areas in the valleys of both the Colorado and Gila rivers and especially in Yuma and Pima Counties. Placer deposits are known along the Colorado from a point south of Yuma to as far north as Ehrenberg. There are some around Quartzsite, La Paz, and twenty miles south of Yucca in the Chemehuevis district of Mohave County. Other areas are situated in Yavapai County in the Weaver, Walker, and Big Bug districts and along the Hassayampa River south into the San Domingo district of Maricopa County. Gold placers are also found at Greaterville, Quijotoa, and north of Arivaca in Pima County, in the Teviston and Dos Cabezos districts in Cochise County, at Oracle in Pinal County, and deposits of minor importance exist in Santa Cruz, Greenlee, and Pima Counties. The locations on the accompanying map are only approximate, but serve to show the districts described in detail later.

PIMA COUNTY

GREATERVILLE PLACERS (9)

The information here given concerning the Greaterville placers has been taken from a complete and detailed description of these deposits given in United States Geological Survey Bulletin No. 430 by J. M. Hill, and from data secured by the writer when he visited the area in 1919 and 1922.

These placers, occurring for the most part in an area with the town of Greaterville at the center, are situated on the eastern slope of the Santa Rita Mountains about thirteen miles northwest by road from the town of Sonoita on the Nogales branch of the Southern Pacific Railroad.

The placer area is flat country deeply and rather precipitously cut by arroyas which drain towards Cienega Creek. There is no surface water except in the rainy seasons, when all the principal gulches are filled. These streams together with wells supply the local need, but there is hardly enough water to carry on rocker work to advantage.

Briefly, the principal rocks of the district may be said to comprise granite, which is intensely weathered, silicified slates, sandstones, and wash material that covers the lower slopes of the mountains and the valley floor. The older granite, slates, and sandstones are intruded by granite porphyry and narrow rhyolite dikes. Vein deposits occur, and are directly

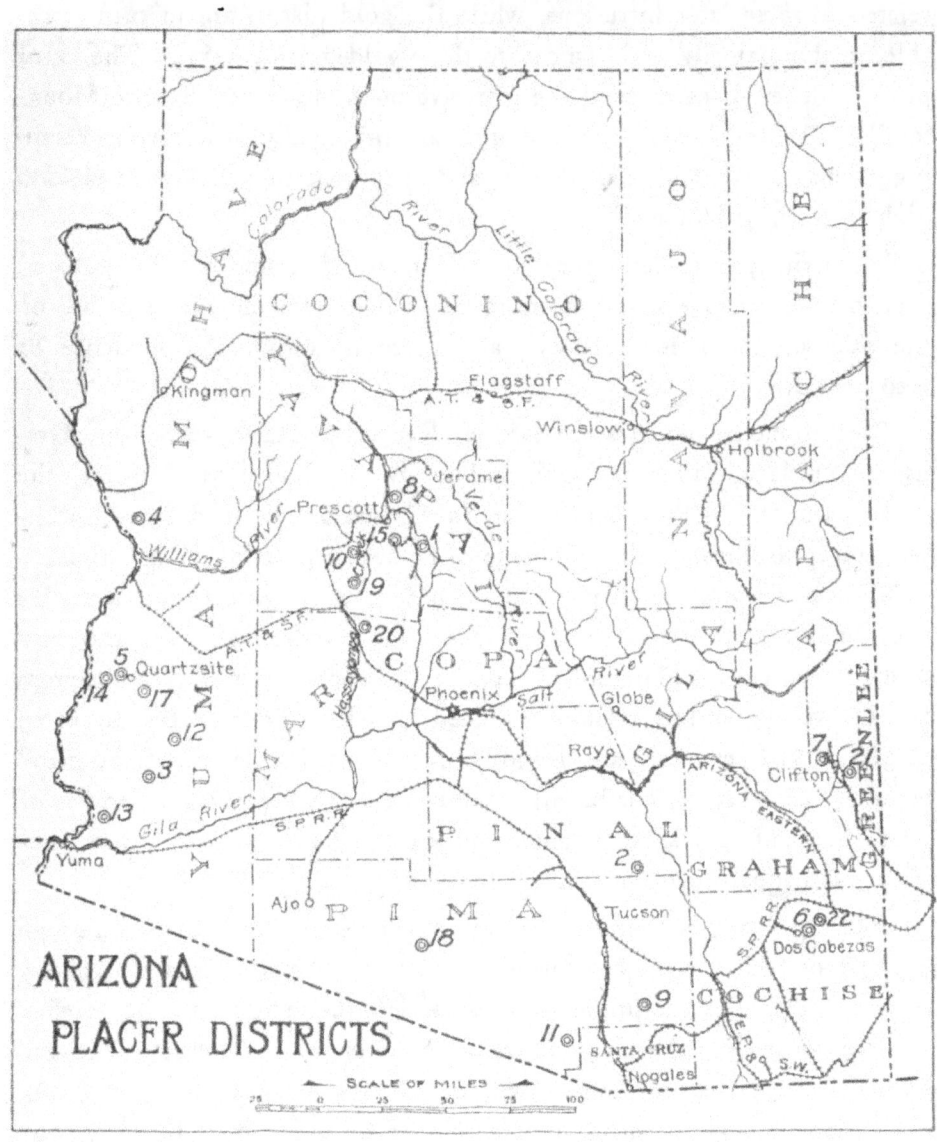

ARIZONA
PLACER DISTRICTS

1.	Big Bug	12.	Kofa
2.	Canada del Oro or Old Hat	13.	Laguna
3.	Castle Dome	14.	La Paz
4.	Chemehuevis	15.	Lynx Creek
5.	Dome Rock	2.	Old Hat or Canada del Oro
6.	Dos Cabezas	17.	Plomosa
7.	Gold Gulch	18.	Quijotoa
8.	Granite Creek	19.	Rich Hill
9.	Greaterville	20.	San Domingo
10.	Groom Creek	21.	San Francisco
11.	Las Guijas	22.	Teviston

related to these later intrusions, while the gold placers are in turn traceable, in the majority of instances, to these gold-bearing veins. The chief placer gulches all head about the intrusive mass known as "Granite Mountain". The vein deposits of the areas are principally quartz veins carrying galena, pyrite, and chalcopyrite, and at the outcrops they have yielded high values in gold and silver.

The principal placers were found in the bottoms of the gulches, though older channels have been found which cross the ridges or are on the sides of the present valleys, and there is considerable evidence of reconcentration.

The productive gulches are Boston, Kentucky, Harshaw, Sucker, Graham, Louisiana, Hughes, Ophir below its junction with Hughes, the upper parts of Los Pozos and Colorado, Chispa on the road from Enzenburg camp to Greaterville, and Empire below its junction with Chispa.

The pay channels vary in width from four to fifty feet, but in Sucker and Empire gulches the widths reached one hundred and one thousand feet, respectively. The gold is found at bedrock at depths varying from two to twenty-five feet. Gold produced in the district has been marketed at $16 to $19 per ounce and is mostly fine and flaky. Some of the metal is rusty, but most of it is bright. Under the microscope the particles of gold are rough and angular, and many of the larger nuggets "were associated with quartz and galena."

[1]"The pay dirt is found on bedrock distributed rather evenly through a two-foot bed of angular gravels in a fine red-brown, somewhat clayey matrix . . . . . The conditions were essentially the same in all the gulches, and the thickness of the pay dirt varied little from place to place. The constituents of the bed are rather fine, usually less than an inch in greatest dimension, though in many places cobbles of four to eight inches are found. In a few places the materials of this bed are roughly stratified and somewhat cemented, usually by lime." This latter statement is rather a significant one, and indicates a condition that is frequently met with in Arizona placers. This cemented material is the common desert "caliche", a formation that adds greatly to the difficulty of placer mining in this state. In the Quijotoa district it is so extensive that crushing machinery is necessary to liberate the gold from the richer gravels. This

---

[1]Notes on the Placer Deposits of the Greaterville District, Ariz., by J. M. Hill. United States Geological Survey Bulletin No. 430.

condition also occurs in the placers on the east slope of the Dome Rock Mountains and in the Plomosa placers, both in Yuma County.

Around Greaterville the constituent pebbles are very angular and show almost no water wear. Even the sand consists of angular broken fragments rather than rounded grains. These pebbles are held together by a red-brown clay, not very difficult to handle with water.

"The depth of this bed varied in the different localities, being almost at the surface in the heads of the gulches, and buried to depths of ten to twenty feet in the lower ends of the diggings.

"The Cambrian (?) sedimentary rocks form a perfect bedrock in the upper parts of the gulches. The beds are standing on edge, and their differences in weathering, due to difference in hardness, have formed natural riffles, behind which the gold has been concentrated. In the lower parts of Kentucky, Sucker, Ophir, and Empire gulches the 'cement rock' . . . . . forms the bedrock and its rough surface has acted as riffles. The bedrock in Colorado, Los Pozos, and Louisiana gulches is entirely 'cement rock'. This shows that the concentration of the gold has been at least later than early Quaternary."

It is stated that later investigations in the Greaterville district below this so-called "cement rock" have proven the existence of gold in and below it.

Water for any kind of placer work in the Greaterville district is scarce. All water for working in the earlier days was packed from Gardner canyon to the diggings by burros. Clay in the pay dirt makes hand methods and rocking difficult. Mexicans who at present work the gravels do so by sinking shallow shafts to bedrock where they mine the gold-bearing gravel around these shafts for a considerable radius, hoist it to the surface, and, when sufficient has been mined to run a rocker for several days, purchase water and treat the dirt.

Hill states that two attempts have been made to work these gravels on a large scale; one, soon abandoned, by means of a small steam shovel in Empire Gulch; the other by hydraulicking the gravels of Kentucky Gulch. This attempt was made by the Stetson Company. In 1900 they constructed an eight-mile pipe line to bring water under a 125-foot head, but ceased work after several months of operation.

In 1905 the Santa Rita Water and Mining Company acquired title to a large tract of the ground and commenced operations with extensive equipment, including ten miles of ditch and pipe line. The water for the operations was obtained by impounding dams in the canyons. Hydraulic operations were carried on during the winter months and early spring of 1905. Lack of ample storage capacity for water made operation in the dry season impossible, and work was discontinued.

In October, 1914, it was reported that the Greaterville Dredge Gold Mining Company had acquired 1100 acres of placer ground which had been thoroughly prospected. Much of the ground was reported to average 90c per cubic yard, and the proposed dredge was to have a capacity of two thousand cubic yards per day. Water to carry on the dredging operations was to have been obtained from wells on the property. These plans were never consummated.

At the time this Bulletin was prepared an Eastern concern was investigating the possibility of working these placers profitably by hydraulic methods.

The total actual production to date has been estimated at $7,000,000 by Mr. J. P. Coyne. This figure, according to Hill, was corroborated by several old time miners of the district. Hill further states that E. Ezekial, a mining engineer who had investigated the Greaterville placers, places the quantity of gold still remaining in an area of about eight square miles at $100,000,000.

The Las Guijas Placers (11)

The Las Guijas placers lie in the northern, northeastern, and southern slopes of the Las Guijas Mountains, a range about sixty miles south of Tucson and twenty miles due west of Amado station on the Nogales branch of the Southern Pacific Railroad. This range is small and low, but rises abruptly from the mesa lands.

On the southern slope the area between Guijas Creek and the Guijas Mountains for a distance of two and a half miles is reported to be covered with gold placer gravel.[1] This gravel extends up the gulches into the mountains themselves. On the southern slopes of the mountains there is no mesa gravel, but placer gravel occurs in the arroyas.

This gold originated in the numerous gold-bearing veins and stringers of the Guijas Mountains from which it has been washed on the mesas and into the arroyas. The mesa gravels vary in thickness from a mere

---

[1]Professional Report of Las Guijas Placers, by I. C. Friend.

layer to sixteen feet. Their gold is evenly disseminated, occurs rough and unpolished, and is frequently still attached to the original gangue minerals. The arroya gold is more rounded and polished. A cemented clay layer occurs between the richer pay dirt and bedrock. There appears to be as much placer gravel on the southern slopes as on the northern, except that on the former the gravel is more patchy, but thicker and richer due to greater concentration.

For the past fifty years these placers have been worked by Mexicans and Indians, who use the rocker in the rainy seasons and the batea in dry periods. There are many evidences that extensive work has been done here, and it is claimed that sufficient gold has been produced to start at least one Mexican revolution. The method at present used is to sink a shaft to bedrock, scrape up the rich two- to four-inch layer there, and treat it for gold.

In 1905 the New Venture Gold Placer Company had taken up forty-two hundred acres, and it was claimed that development work had shown that about 4800 cubic yards of gravel that ran a dollar to the cubic yard existed on each acre of the entire area. The company proposed to pump water for hydraulicking three miles, obtaining it from the Arivaca River, a stream that flows the year round. Water can also be obtained at depths of thirty to forty feet from wells on Guijas Creek.

QUIJOTOA PLACERS (18)

The Quijotoa placer district is located seventy miles almost due west of Tucson, and extends from Quijotoa on the north to the Mexican boundary. The placers carry gold from the surface through the underlying caliche which constitutes the richest part of the deposit. There the gold is coarse, and occurs in the cement that unites the pebbles and rock fragments.

This area has been worked in places by the Indians and Mexicans for hundreds of years, probably by the same methods they now use. They are able to treat only the dirt and the softer patches of caliche, which they handle in a very primitive fashion. The caliche is beaten in rawhide bags, and the gold is recovered from the pulverized cement by crude hand machines. This process is antiquated, yet it is the only one that has been successful, and all attempts made by white miners to work these placers have failed because of the hardness of the caliche and the scarcity of water.

In 1910[1] a Quinner pulverizer that had been used successfully in the Altar District of Sonora, Mexico, and a Stebbens dry concentrator were installed by the Manhattan Company in the Horseshoe basin area of the Quijotoa district, but the experiment proved unprofitable although the yield for the whole district is said to be in the neighborhood of a dollar a cubic yard.

In 1905 and 1906 the United States Geological Survey Mineral Resource publications stated that most of the productive ground was owned by the Imperial Gold Mining Company, and was leased to placer miners.

Ash Creek Placer

A very small placer deposit occurs along Ash Creek in the Papago mining district on the western slope of the Sierrita mountains, about twenty-five miles southwest of Tucson. The creek traverses the Sunshine-Sunrise group of claims and it is there the placer diggings occur.

The area covered by the auriferous gravel is very small, but Mexicans working in the rainy seasons are said to make good wages by the use of rockers. There is ample water in the creek for the use of such apparatus then, and the remains of old diggings indicate that a considerable amount of work has been done there in the past.

YUMA COUNTY

La Paz Placers (14)

The La Paz placers lie on the west slope of the Dome Rock Mountains, in Yuma County, about nine miles northeast of Ehrenberg. The town of La Paz was situated on part of the present Colorado Indian Reservation. It is now nothing but a ruin in spite of the fact that in 1863 there were 1500 people there. The placers lie in Ferra, Garcia, Ravenna, and Goodman gulches due east of the Colorado River. They are reached by road from Quartzsite, a town twenty-five miles south from Bouse station on the Phoenix-Parker branch of the Santa Fe Railroad.

The topography of the region is not precipitous. The desert bench lands slope gently westward from the mountains to the river where there

---

[1] Dry Placer Mining in the Arizona District, by F. W. Fickett.

is a sudden drop to the river-bottom lands. This area, however, is traversed by fairly wide and deep arroyas, which carry off all flood waters in the so-called rainy seasons. The climate is so arid, however, that there is no surface water during most of the year, but Tyson wash at Quartzsite carries a small underground flow. Indeed, water is so scarce at the La Paz placers that it has to be hauled in for camp purposes.

The geology of the La Paz placer area, according to E. L. Jones, Jr.,[1] consists of a quartz-epidote schist, an altered igneous rock that has been intruded by a very much younger granite. This schist is the rock that contains the gold-bearing quartz veins and stringers from which the gold of the placers has been derived.

The following description of the La Paz placers has been taken from United States Geological Survey bulletin 620-C, Gold Deposits near Quartzsite, Ariz., by E. L. Jones, Jr.

"*Character of Gold-Bearing Wash.* The gold-bearing material consists of sand and clay inclosing angular rock fragments of greatly variable size. Tests indicate that about twenty percent of the wash will pass through a quarter-inch screen, and the largest boulders weigh several hundred pounds. The material near the surface is unassorted and is unconsolidated, being readily worked with pick and shovel. That at depths of fifteen or twenty feet is consolidated, but the cementing substance readily disintegrates on exposure to air. Deposits of wash below the depths of test pits may prove to be similar to the outwash on the east slope of the Dome Rock Mountains and in the Plomosa placers, where the material is firmly cemented with calcium carbonate and requires crushing in order to free the gold. In Goodman Wash below the Goodman tank a deposit of calcareous tufa several feet thick was noted. The ground stands sufficiently well to permit the sinking of shafts without the use of timber. The wash is readily worked in dry-washer machines, the only requirement being that the ground must be dry. The gold is said to be distributed throughout the wash, though in the early workings the richest yield was obtained near bedrock. The size of the gold now recovered from the deposits of the La Paz district probably averages only a few cents, but, as already stated, the gold recovered from the early workings was much coarser. The gold is rough and angular, and particles of iron cling to

[1]United States Geological Survey Bulletin 620-C, Gold Deposits near Quartzsite, Arizona, by E. L. Jones, Jr.

some of the nuggets. Magnetite is always found in the concentrates, and boulders of magnetite, the largest weighing several pounds, are frequently found on the surface."

The richer gravels of the La Paz placers were worked over many years ago, but only the coarser gold was extracted then since the miners were handicapped by lack of water, and they used only pans, rockers, and dry washers to treat the gold-bearing ground.

Several years ago the New La Paz Gold Mining Company acquired by location and purchase all the placer ground south of the Colorado Indian Reservation. The dividing line was to follow the Arroya La Paz. This company proposes to wash the placer gravels by pumping water from the Colorado River in sufficient quantity to hydraulic and sluice three thousand cubic yards of gravel a day. The water will be pumped through pipe lines for a distance of four and one-half miles to a reservoir at an elevation of 540 feet above the river bottom. This will give a head of 225 feet for washing the gravels. Engineers, reporting on the property for the company, have estimated, from samples taken in seventy-four shafts, varying in depths from five to forty-five feet, that there are 1,300,000 cubic yards that will average $1.80 per cubic yard. These engineers also state that the gravel is free from clay, is uncemented, and contains but few boulders. Part of the machinery to carry out this undertaking was installed several years ago, but the work has been held up by litigation with the Interior Department over the true location of the southern boundary of the Colorado Indian Reservation. This dispute has been settled, however.

The consensus of opinion of many of the older miners, who have a knowledge of the earlier day placer operations in the La Paz district, is that the value of production was about $7,000,000 from 1862 to 1868. Since that date operations and production have been spasmodic for several reasons: the gravels have already been worked over and the coarser gold recovered; without a large investment only dry washers can be used, and such machines are not adapted to the treatment of large quantities of gravel; and, finally, the ground itself varies in richness.

PLOMOSA DISTRICT PLACERS (17)

The following description of the placers on the east slope of the Dome Rock Mountains has been taken from the United States Geological

Survey Mineral Resources for 1912, Arizona Dry Placers, by V. C. Heikes:

"The Plomosa mining district, lying east of the Colorado River in Yuma County, Ariz., is situated in the Posas Valley, a great north and south depression, with the Plomosa Mountains forming the eastern border and the northward extension of the Castle Dome Range on the west. Posas Valley trends northward thirty or forty miles, and is from ten to fifteen miles wide. It slopes to the north and affords a very extensive field . . . . In this valley are situated . . . . on its eastern side the Plomosa placer ground and on the western side an extended deposit of gold-bearing gravel known as the La Cholla, Oro Fino, and Middle Camps. In some localities pits have been sunk to a depth of twenty, thirty, and fifty feet or more to beds of cement which are richer than the gravel. Near the mountain the gold is coarser, but the gravel is much less. Miles of the great deposit extending westward from the mountains, and from three to four miles in width, have been cut into by floods from the mountains, forming deep ravines, and they afford miles of banks ten to fifteen feet high in which the upper layer of gravel is well exposed. From these banks, as far as investigations could be made, samples gave an average return value of 64c per cubic yard with gold estimated at $18 per ounce. . . . . . There were no failures. The results lay between the extremes of 42c and $1.04 per cubic yard. To get the limit of the deposit it would be necessary to pursue the tests to points where gold failed . . . . . A dry washer was used, winnowing the sand by means of a rotary fan, which blows under a fine screen over which the sand passes in a thin stream . . . . . The limit of the gravel actually explored was 2,400 by 1,500 feet and eight yards deep . . . . . Within this area bedrock was not reached at any time . . . . . "

DOME ROCK PLACERS (5)

"Across the valley and about twelve miles from the Plomosa placer is another bed of gravel of similar extent and of high quality . . . . . This western mass of gravel is occupied at various places by three camps, all rich in gold, and all differing materially in character of gravel. Middle Camp, the most northerly of the three, has granite gravel; Oro Fino, in the center, has much porphyritic slate; and La Cholla, at the south, is mostly composed of quartzite and schist pebbles . . . . . At La Cholla

. . . . . , which is nearer the mountains, there is a siliceous cement, very rich, but also so very hard that it requires to be broken by powder before going to the dry washer. At Oro Fino the shale bedrock is very near the surface. In Middle Camp there is cement, but of a much softer kind . . . . . (Here most of the sampling was performed.) . . . . . The camp (?) occupies the east and west valley crossing the mountain range, a mile wide and four or five miles long . . . . . This is the chosen locality for the individual dry washer, who takes his machine to some point where the bedrock can be reached quickly. It is here that the rich seams of gravel on the bedrock yield from four to ten times the value of the thicker gravels, and in crevices there have been found nuggets worth $10 to $25. La Cholla, south of Middle Camp, lies along the foot of the mountains, like Plomosa, and is three or four miles in length . . . . . The depth of the gravel is irregular in passing from Middle Camp through Oro Fino to La Cholla. Forty years ago the Colorado River was the main gateway, and freight for many years entered that way. The value of these placers was known to the miners who, in that early day, passed over all the region adjoining the Colorado, but the almost total absence of water in the mountains compelled the miners to pack their rich dirt to the river or to distant tanks to be washed. Oro Fino was the most celebrated camp of that day. There the soft shale bedrock rises to the surface; and when the art of dry washing was learned the rich bedrock was the scene of active work.' "

"Surrounding the post office of Quartzsite, in the Plomosa mining district, and extending in every direction, covering an area of about 7,500 acres, is found dry-placer ground with values to an average depth of fifteen feet and varying from five to fifty feet. The coarse gold content per cubic yard is reported to average from 10c to several dollars. Efforts are being made to get a combination of equipment which will successfully work the desert gravels and gold-carrying cement gravels at a minimum cost per yard. At a point south of Quartzsite and seventeen miles north of Ehrenberg, on Colorado River, a reduction plant capable of handling two thousand cubic yards of gravel every twenty-four hours is reported under construction. The machinery consists of a Barnett dredger, a Quenner disintegrating machine, and a Stebbins concentrator."

Kofa Placers (12)

According to Jones,[1] the placers of the Kofa Mountains in Yuma County occur in a gulch that drains the hills in which the King of Arizona Mine is located. The gravels are said to cover an area of sixty acres, and range from a few feet to seventy feet in thickness. These gravels are the eroded products of the metamorphic and volcanic rocks of the area. The gold which they contain is coarse, and occurs near bedrock.

The production of the district is said to be about $40,000 to date. The deposits are still worked in a crude and small way, and yield a few hundred dollars annually.

YAVAPAI COUNTY

Walker District, Lynx Creek Placers (15)

The Lynx Creek placers are located along Lynx Creek, five and one-half miles due east of Prescott, and extend for a distance of five miles from Walker to Lonesome Valley. According to Hamilton[2] the Lynx Creek placers were among the richest in Arizona, and he states that it was estimated that they yielded $1,000,000 prior to 1881.

Operations are still carried on spasmodically along Lynx Creek and its tributaries when water is available. A small amount of gold, chiefly coarse, is recovered by the use of rockers and dry washing machines. Several large scale operations have, however, been attempted. In 1907-8-9 a dredge was installed and operated, and a small amount of gold was recovered, but the enterprise was apparently unprofitable, and was abandoned. The last effort made to work these placers by machinery was in 1918-19, when a drag-line scraper, screens, and a washer were installed. Operations were soon stopped, however, and the plant deteriorated and was finally removed.

On Lynx Creek as far up as Walker there are old debris piles, the remnants of early day ground sluicing operations. They consist mostly of coarse boulders.

---

[1]United States Geological Survey Bulletin 620, Reconnaissance of the Kofa Mountains, by E. I. Jones, Jr., pp. 151-164.
[2]Resources of Arizona, by Patrick Hamilton, 1881.

Rich Hill and Other Weaver Placers (19)

The Weaver District placers are located along Weaver Creek and at Rich Hill, about 12 miles southeast of Congress Junction. The Weaver Creek deposits were discovered by Captain Pauline Weaver in 1863 and were the first placers worked in Yavapai County. The placers of Rich Hill in the same district were discovered by a Mexican who was on his way to the Weaver camp. They were found in a depression on the summit of a mountain, and the coarse gold was lying bare on the bedrock in many places. Large pieces of gold worth hundreds of dollars were not uncommon, and metal valued at $5000 to $6000 was frequently found under a single boulder. The gulches and ravines running down from the mountains also contained considerable gold. $500,000 is said to have been taken from an acre of ground, and Hamilton[1] states that $1,000,000 was yielded by these placers prior to 1883. This figure is probably low. During and following rainy seasons these placers are still worked by numerous miners, and the accompanying table gives the value of the production for the last twenty years.

The placer ground that contains the gold extends over an area of five by eight miles, and is ten feet or more deep. The gold in the stream gravels is coarse, but that in the soils and gravels of the mesa is fine.

Groom Creek Placers (10)

The placers along Groom Creek, ten to fifteen miles southwest of Prescott, have been worked over for the sixth time according to Sparks,[2] and have produced $3,000,000 in gold.

Hassayampa Placers

Gold placer gravels occur practically all along Hassayampa Creek from the Yavapai-Maricopa County line north up the stream to its source near Walker, as well as along its many small tributaries, such as Copper Creek, Ash Creek, Slate Creek, East Hassayampa Creek, Cherry Creek, and Oak Creek.

These placers are still worked during and following the rainy seasons, when small amounts of gold are recovered. In the dry seasons dry placer machines are occasionally used. Attempts at large scale operations have

[1] Resources of Arizona, by Patrick Hamilton, 1883.
[2] Yavapai, The Land of Opportunity, by G. M. Sparks, Arizona Bureau of Mines Bulletin No. 59.

been made from time to time, but have invariably met with failure. Jagger and Palache[1] state that "it has been found that some of the gravely beds in the western belt of volcanic agglomerate are auriferous, ... and on Slate and Milk Creeks some hydraulic washing is being done on deposits belonging to this formation."

OTHER PLACERS

Other placer areas in Yavapai County are found along Castle Creek, Humbug Creek, Turkey Creek, Granite Creek (8), and Big Bug Creek (1). In 1900-01 a dredging plant was erected near Mayer on Big Bug Creek. This enterprise met with failure, due probably to the fact that the ground was not suitable for dredging, because of the great number of very large boulders. An attempt was made to work the placers on Humbug Creek in the early nineties by an English company. This company spent a considerable amount of money in building a camp, and in constructing pipe lines and dams to carry on hydraulicking. The attempt failed through lack of water and inability to save the gold known to be present.

COCHISE COUNTY

DOS CABEZAS PLACERS (6)

The Dos Cabezas placer field was discovered in 1901 by some Mexican prospectors. They recovered only a small amount of gold, but enough was obtained to create considerable local excitement. All the streams and gulches to the north and south of the town were soon being worked. These gravels are comparatively shallow except at the town of Dos Cabezas itself. At no small distance north into the canyons the amount of gravel becomes negligible. The gold comes from the numerous gold quartz veins and stringers that are present in the six-mile belt of Mesozoic sediments that occur along the south edge of the Dos Cabezas mountains. Water was very plentifiul in 1906, so a considerable amount of placer work was then done.

TEVISTON PLACERS (22)

The Teviston placers, located on the north side of the Dos Cabezas

---

[1]United States Geolrgical Survey Bradshaw Mountain Folio, No. 126, by T. A. Jagger and Charles Palache.

Mountains, opposite the Dos Cabezas district, have yielded some gold when they have been worked in the wet season by dry placer methods. According to Heikes[1], "Some three hundred acres have been reported as being valuable to a depth of three to ten feet. Bedrock is fifty to seventy feet deep. Most of the gold is coarse, and the ground by tests has yielded from 3c to $28 per cubic yard. The largest nugget found was valued at $375. Some cement or caliche has been found in prospecting the ground, but values have been found in the gravels beneath."

PINAL COUNTY

Canada del Oro or Old Hat Placers (2)

Captain J. D. Burgess in some communications to Heikes[2] describes the placers in the Old Hat district and says that in "an area of 25,000 acres, ... covering nearly the whole of Tp. 10 S., R. 14 E., Gila and Salt River meridian, distant four to ten miles south from Oracle post-office and sixteen to twenty-nine miles north from Tucson, is found valuable dry placer gravel, which has apparently been deposited at intervals by floods from the Santa Catalina Mountains so as to form a deposit of nearly equal value from surface to bedrock, there being no pronounced accumulation of heavy gold at bedrock except in the stream, Canada del Oro Creek, which passes through the region. The bed of dry gravel is from six feet deep at the creek side to 252 feet at the summit, with an average thickness of about 150 feet. The deposit is in general a loose gravel, uncemented. There are, however, alternating strata of deep red, clayey material. These strata are of nearly uniform thickness of three to four inches and probably were formerly surfaces existing between floods, each being covered by a later flow of gravel from rainfall eroded veins farther up the mountain. Shafts sunk on the hillsides from twenty-seven to fifty feet in depth show values from 10c to 42c per cubic yard. The average is difficult to determine, as the gold is not equally distributed. All the gold is found in well rounded nuggets ranging from 50c to $5.00 in value. There is a tradition that a lump weighing 16 pounds, probably 40% of which was quartz, was once found, but its discoverers were found murdered in their camp sixteen miles north of Tucson. The nugget had

---

[1]United States Geological Survey Mineral Resources for 1912. Arizona Dry Placers, by V. C. Heikes, pp. 255-260.
[2]Idem.

disappeared. In fineness the gold averages about 0.905. Generally the placer material is dug, screened, and hauled to the creek, and there worked by rockers, or sluiced when there is enough water. Many dry washers have been tried, but most of the gold lies in the red clayey seams which apparently acted as 'bedrock' for each period of deposition. Pulverizing this adherent material gives good results with the common bellows type of 'dry wash'. A boiler and pump were once used to throw water against the creek bank, but the water at that time proved insufficient for extensive operations."

MARICOPA COUNTY

SAN DOMINGO PLACERS (20)

Placer gravels occur along the Hassayampa and its tributaries at a point forty-five to fifty miles northwest of Phoenix and near Morristown on the Santa Fe Railroad. Gold has been known to occur in this district, the San Domingo, for many years. Most of the country is rolling. The gold in the placers is derived from gold quartz veins in schists, gneisses, and granites which have been intruded by Tertiary andesites, rhyolites, etc.

The gold-bearing gravels are found in numerous gulches that lead down from the hills to the northeast of Morristown. During heavy rains these gulches carry the run-off to the Hassayampa, and into them is washed the gold-bearing sand and gravel. The coarse gold is found at bedrock in the upper reaches of the gulches, but here the amount of gravel is small. In the wider washes, according to T. Lane Carter,[1] are found placer gravels that cover areas as much as twelve hundred feet wide and two and one-half miles long. The average depth to bedrock is ten feet, and the average value, as determined from samples taken from test pits to bedrock, is 40c per cubic yard. In the larger deposits, and along the lower reaches of the gulches, the gold is distributed fairly uniformly.

One property in this district, called the Lotowana Mine, and located by John Sanger, covers four thousand acres. One-tenth of this area has been tested by pits. Part of the property, on "Rogers Wash," is said to be covered with a gravel bed that is two and one-half miles long, and one thousand feet wide, and one to twenty feet thick. It is estimated to

---

[1]Mining and Scientific Press, August 10, 1912. Vol. 105, Gold Placers of Arizona, by T. Lane Carter.

contain 4,400,000 cubic yards that will yield 40c per cubic yard. Some of the ground will average 80c per cubic yard, however.

Several engineers have proposed to build storage dams to catch the heavy run-off water, and to use it in hydraulicking this area. It has also been suggested that the underflow of Hassayampa Creek could be caught and diverted by flume and pipe line, and used for hydraulic mining. The particles of gold in these deposits are angular, the material is sandy with few big boulders, and clay and caliche are said to be absent.

The value of the production from this district to date is not known, but the accompanying table gives the amount of gold produced since 1900.

In 1908 material was hauled for the construction of a dredge on the Agua Fria River in Maricopa County, but nothing is known of the results obtained.

MOHAVE COUNTY

In 1909 a suction type dredge to treat the gold sands of the Colorado River was constructed on the Arizona side of the river opposite Eldorado Canyon, Nevada. The capacity of the dredge was seven thousand cubic yards daily. It failed in the first test to extract the fine gold, and was shut down. Subsequently, in the spring of 1910, it was torn from its moorings at high water and wrecked.

Recently there has been considerable excitement over the possibility of the existence of rich placer gravel along Silver Creek below Oatman. Test pits are being sunk to bedrock. What success this new undertaking has met with has not yet been divulged.

Placer gold gravels that have yielded some gold occur in the Chemehuevis district (4). This district is located twelve miles south of Franconia on the main line of the Santa Fe railroad.

GREENLEE COUNTY

According to Lindgren[1] "placer gold has been found in Gold Gulch(7), as fine flakes in the gravels along the Morenci Canyon about four miles below the town, and in several places in gravels along the San Francisco River (21). The best prospect occurs near Oroville, where work-

[1]United States Geological Survey Professional Paper No. 43. The Copper Deposits of the Clifton-Morenci District, Ariz., by Waldemar Lindgren.

.ing has been attempted. . . . . . The gravels lying in front of the hills of the older rocks at Morenci and Clifton are auriferous in places. Placers of some value were worked in Gold Gulch, but are now exhausted. An unsuccessful attempt was made some years ago to mine, by the hydraulic method, the bench gravels of the San Francisco River, which doubtless derived their gold from the veins northeast of Copper Mountain. The Gila conglomerate south of Morenci contains fine gold which is concentrated in shallow gullies. Payable placers have not been found."

DRY PLACER MACHINES

Both hand and power machinery has been devised to treat the dry placer gravels of Arizona. Most of these machines have been operated at one time or another on Arizona placers. For a complete discription of these machines, the reader is referred to the literature on this subject which is listed in the bibliography of this bulletin.

BIBLIOGRAPHY

United States Geological Survey Bulletin No. 430. Notes on Placer Deposits of Greaterville, Arizona, by J. M. Hill.

United States Geological Survey Bulletin No. 451. Reconnaissance of the Ore Deposits in Northern Yuma County, by Howland Bancroft. La Paz and Plomosa Placers, pp. 85-88.

United States Geological Survey Bulletin No. 582. Mineral Deposits of the Santa Rita and Patagonia Mountains, by F. C. Schrader. Greaterville placers, pp. 160-166.

United States Geological Survey Bulletin 620. Gold Deposits near Quartzsite, Arizona, by E. L. Jones, Jr., pp. 45-57.

United States Geological Survey Bulletin 620, Reconnaissance in the Kofa Mountains, Ariz., by E. L. Jones, Jr., pp. 151-164.

United States Geological Survey Mineral Resources for 1900 to 1920.

United States Geological Survey Mineral Resources for 1912, Arizona Dry Placers, by V. C. Heikes, pp. 255-260.

Report on the Mineral Resources of the States and Territories west of the Rocky Mountains, by J. Ross Browne. 1868, pp. 443-482.

Mines and Mining West of the Rocky Mountains, by R. W. Raymond. 1875, p. 390 and 1876, p. 342.

Production of the Precious Metals in the United States. 1884, p. 46, by H. C. Burchard.

United States Geological Survey Folio No. 126, Bradshaw Mountains, by Jagger and Palache.

Report of the Governor of Arizona for 1899, Mining in Arizona, by Prof. W. P. Blake.

Arizona the Land of Sunshine and Silver, by J. A. Black. 1890.

Mining and Scientific Press, August 10, 1912, Vol. 105. Gold Placers of Arizona, by T. Lane Carter.

Arizona Magazine, November, 1915. Las Guijas Placers, by Charles F. Willis.

Mining and Scientific Press, Vol. 102, 1911, p, 291, Dry Placer Mining in Quijotoa District, by F. W. Fickett.

Resources of Arizona, by Patrick Hamilton, 1883.

BIBLIOGRAPHY OF DRY PLACER MACHINES

Dry Pulverizer and Separator. Engineering and Mining Journal, Vol. 89, p. 858.

Mining and Scientific Press, Vol. 102, 1911, p. 291. Dry Placer Mining in Quijotoa District, by F. W. Fickett.

Mining and Scientific Press, Vol. 105, 1912, pp. 50-52. Dry Concentration of Placer Gold, by F. J. H. Merrill.

GOLD RUSH BOOKS

OREGON, USA

www.GoldMiningBooks.com

Books On Mining

Visit: www.goldminingbooks.com to order your copies or ask your favorite book seller to offer them.

Mining Books by Kerby Jackson

Gold Dust: Stories From Oregon's Mining Years - Oregon mining historian and prospector, Kerby Jackson, brings you a treasure trove of seventeen stories on Southern Oregon's rich history of gold prospecting, the prospectors and their discoveries, and the breathtaking areas they settled in and made homes. 5" X 8", 98 ppgs. Retail Price: $11.99

The Golden Trail: More Stories From Oregon's Mining Years - In his follow-up to "Gold Dust: Stories of Oregon's Mining Years", this time around, Jackson brings us twelve tales from Oregon's Gold Rush, including the story about the first gold strike on Canyon Creek in Grant County, about the old timers who found gold by the pail full at the Victor Mine near Galice, how Iradel Bray discovered a rich ledge of gold on the Coquille River during the height of the Rogue River War, a tale of two elderly miners on the hunt for a lost mine in the Cascade Mountains, details about the discovery of the famous Armstrong Nugget and others. 5" X 8", 70 ppgs. Retail Price: $10.99

Oregon Mining Books

Geology and Mineral Resources of Josephine County, Oregon - Unavailable since the 1970's, this important publication was originally compiled by the Oregon Department of Geology and Mineral Industries and includes important details on the economic geology and mineral resources of this important mining area in South Western Oregon. Included are notes on the history, geology and development of important mines, as well as insights into the mining of gold, copper, nickel, limestone, chromium and other minerals found in large quantities in Josephine County, Oregon. 8.5" X 11", 54 ppgs. Retail Price: $9.99

Mines and Prospects of the Mount Reuben Mining District - Unavailable since 1947, this important publication was originally compiled by geologist Elton Youngberg of the Oregon Department of Geology and Mineral Industries and includes detailed descriptions, histories and the geology of the Mount Reuben Mining District in Josephine County, Oregon. Included are notes on the history, geology, development and assay statistics, as well as underground maps of all the major mines and prospects in the vicinity of this much neglected mining district. 8.5" X 11", 48 ppgs. Retail Price: $9.99

The Granite Mining District - Notes on the history, geology and development of important mines in the well known Granite Mining District which is located in Grant County, Oregon. Some of the mines discussed include the Ajax, Blue Ribbon, Buffalo, Continental, Cougar-Independence, Magnolia, New York, Standard and the Tillicum. Also included are many rare maps pertaining to the mines in the area. 8.5" X 11", 48 ppgs. Retail Price: $9.99

Ore Deposits of the Takilma and Waldo Mining Districts of Josephine County, Oregon - The Waldo and Takilma mining districts are most notable for the fact that the earliest large scale mining of placer gold and copper in Oregon took place in these two areas. Included are details about some of the earliest large gold mines in the state such as the Llano de Oro, High Gravel, Cameron, Platerica, Deep Gravel and others, as well as copper mines such as the famous Queen of Bronze mine, the Waldo, Lily and Cowboy mines. This volume also includes six maps and 20 original illustrations. 8.5" X 11", 74 ppgs. Retail Price: $9.99

Metal Mines of Douglas, Coos and Curry Counties, Oregon - Oregon mining historian Kerby Jackson introduces us to a classic work on Oregon's mining history in this important re-issue of Bulletin 14C Volume 1, otherwise known as the Douglas, Coos & Curry Counties, Oregon Metal Mines Handbook. Unavailable since 1940, this important publication was originally compiled by the Oregon Department of Geology and Mineral Industries includes detailed descriptions, histories and the geology of over 250 metallic mineral mines and prospects in this rugged area of South West Oregon. 8.5" X 11", 158 ppgs. Retail Price: $19.99

Metal Mines of Jackson County, Oregon - Unavailable since 1943, this important publication was originally compiled by the Oregon Department of Geology and Mineral Industries includes detailed descriptions, histories and the geology of over 450 metallic mineral mines and prospects in Jackson County, Oregon. Included are such famous gold mining areas as Gold Hill, Jacksonville, Sterling and the Upper Applegate. **8.5″ X 11″, 220 ppgs. Retail Price: $24.99**

Metal Mines of Josephine County, Oregon - Oregon mining historian Kerby Jackson introduces us to a classic work on Oregon's mining history in this important re-issue of Bulletin 14C, otherwise known as the Josephine County, Oregon Metal Mines Handbook. Unavailable since 1952, this important publication was originally compiled by the Oregon Department of Geology and Mineral Industries includes detailed descriptions, histories and the geology of over 500 metallic mineral mines and prospects in Josephine County, Oregon. **8.5″ X 11″, 250 ppgs. Retail Price: $24.99**

Metal Mines of North East Oregon - Oregon mining historian Kerby Jackson introduces us to a classic work on Oregon's mining history in this important re-issue of Bulletin 14A and 14B, otherwise known as the North East Oregon Metal Mines Handbook. Unavailable since 1941, this important publication was originally compiled by the Oregon Department of Geology and Mineral Industries and includes detailed descriptions, histories and the geology of over 750 metallic mineral mines and prospects in North Eastern Oregon. **8.5″ X 11″, 310 ppgs. Retail Price: $29.99**

Metal Mines of North West Oregon - Oregon mining historian Kerby Jackson introduces us to a classic work on Oregon's mining history in this important re-issue of Bulletin 14D, otherwise known as the North West Oregon Metal Mines Handbook. Unavailable since 1951, this important publication was originally compiled by the Oregon Department of Geology and Mineral Industries and includes detailed descriptions, histories and the geology of over 250 metallic mineral mines and prospects in North Western Oregon. **8.5″ X 11″, 182 ppgs. Retail Price: $19.99**

Mines and Prospects of Oregon - Mining historian Kerby Jackson introduces us to a classic mining work by the Oregon Bureau of Mines in this important re-issue of The Handbook of Mines and Prospects of Oregon. Unavailable since 1916, this publication includes important insights into hundreds of gold, silver, copper, coal, limestone and other mines that operated in the State of Oregon around the turn of the 19th Century. Included are not only geological details on early mines throughout Oregon, but also insights into their history, production, locations and in some cases, also included are rare maps of their underground workings. **8.5″ X 11″, 314 ppgs. Retail Price: $24.99**

Lode Gold of the Klamath Mountains of Northern California and South West Oregon
(See California Mining Books)

Mineral Resources of South West Oregon - Unavailable since 1914, this publication includes important insights into dozens of mines that once operated in South West Oregon, including the famous gold fields of Josephine and Jackson Counties, as well as the Coal Mines of Coos County. Included are not only geological details on early mines throughout South West Oregon, but also insights into their history, production and locations. **8.5″ X 11″, 154 ppgs. Retail Price: $11.99**

Chromite Mining in The Klamath Mountains of California and Oregon
(See California Mining Books)

Southern Oregon Mineral Wealth - Unavailable since 1904, this rare publication provides a unique snapshot into the mines that were operating in the area at the time. Included are not only geological details on early mines throughout South West Oregon, but also insights into their history, production and locations. Some of the mining areas include Grave Creek, Greenback, Wolf Creek, Jump Off Joe Creek, Granite Hill, Galice, Mount Reuben, Gold Hill, Galls Creek, Kane Creek, Sardine Creek, Birdseye Creek, Evans Creek, Foots Creek, Jacksonville, Ashland, the Applegate River, Waldo, Kerby and the Illinois River, Althouse and Sucker Creek, as well as insights into local copper mining and other topics. **8.5″ X 11″, 64 ppgs. Retail Price: $8.99**

Geology and Ore Deposits of the Takilma and Waldo Mining Districts - Unavailable since the 1933, this publication was originally compiled by the United States Geological Survey and includes details on gold and copper mining in the Takilma and Waldo Districts of Josephine County, Oregon. The Waldo and Takilma mining districts are most notable for the fact that the earliest large scale mining of placer gold and copper in Oregon took place in these two areas. Included in this report are details about some of the earliest large gold mines in the state such as the Llano de Oro, High Gravel, Cameron, Platerica, Deep Gravel and others, as well as copper mines such as the famous Queen of Bronze mine, the Waldo, Lily and Cowboy mines. In addition to geological examinations, insights are also provided into the production, day to day operations and early histories of these mines, as well as calculations of known mineral reserves in the area. This volume also includes six maps and 20 original illustrations. **8.5″ X 11″, 74 ppgs. Retail Price: $9.99**

Gold Mines of Oregon - Oregon mining historian Kerby Jackson introduces us to a classic work on Oregon's mining history in this important re-issue of Bulletin 61, otherwise known as "Gold and Silver In Oregon". Unavailable since 1968, this important publication was originally compiled by geologists Howard C. Brooks and Len Ramp of the Oregon Department of Geology and Mineral Industries and includes detailed descriptions, histories and the geology of over 450 gold mines Oregon. Included are notes on the history, geology and gold production statistics of all the major mining areas in Oregon including the Klamath Mountains, the Blue Mountains and the North Cascades. While gold is where you find it, as every miner knows, the path to success is to prospect for gold where it was previously found. **8.5" X 11", 344 ppgs. Retail Price: $24.99**

Mines and Mineral Resources of Curry County Oregon - Originally published in 1916, this important publication on Oregon Mining has not been available for nearly a century. Included are rare insights into the history, production and locations of dozens of gold mines in Curry County, Oregon, as well as detailed information on important Oregon mining districts in that area such as those at Agness, Bald Face Creek, Mule Creek, Boulder Creek, China Diggings, Collier Creek, Elk River, Gold Beach, Rock Creek, Sixes River and elsewhere. Particular attention is especially paid to the famous beach gold deposits of this portion of the Oregon Coast. **8.5" X 11", 140 ppgs. Retail Price: $11.99**

Chromite Mining in South West Oregon - Originally published in 1961, this important publication on Oregon Mining has not been available for nearly a century. Included are rare insights into the history, production and locations of nearly 300 chromite mines in South Western Oregon. **8.5" X 11", 184 ppgs. Retail Price: $14.99**

Mineral Resources of Douglas County Oregon - Originally published in 1972, this important publication on Oregon Mining has not been available for nearly forty years. Included are rare insights into the geology, history, production and locations of numerous gold mines and other mining properties in Douglas County, Oregon. **8.5" X 11", 124 ppgs. Retail Price: $11.99**

Mineral Resources of Coos County Oregon - Originally published in 1972, this important publication on Oregon Mining has not been available for nearly forty years. Included are rare insights into the geology, history, production and locations of numerous gold mines and other mining properties in Coos County, Oregon. **8.5" X 11", 100 ppgs. Retail Price: $11.99**

Mineral Resources of Lane County Oregon - Originally published in 1938, this important publication on Oregon Mining has not been available for nearly seventy five years. Included are extremely rare insights into the geology and mines of Lane County, Oregon, in particular in the Bohemia, Blue River, Oakridge, Black Butte and Winberry Mining Districts. **8.5" X 11", 82 ppgs. Retail Price: $9.99**

Mineral Resources of the Upper Chetco River of Oregon: Including the Kalmiopsis Wilderness - Originally published in 1975, this important publication on Oregon Mining has not been available for nearly forty years. Withdrawn under the 1872 Mining Act since 1984, real insight into the minerals resources and mines of the Upper Chetco River has long been unavailable due to the remoteness of the area. Despite this, the decades of battle between property owners and environmental extremists over the last private mining inholding in the area has continued to pique the interest of those interested in mining and other forms of natural resource use. Gold mining began in the area in the 1850's and has a rich history in this geographic area, even if the facts surrounding it are little known. Included are twenty two rare photographs, as well as insights into the Becca and Morning Mine, the Emmly Mine (also known as Emily Camp), the Frazier Mine, the Golden Dream or Higgins Mine, Hustis Mine, Peck Mine and others. **8.5" X 11", 64 ppgs. Retail Price: $8.99**

Gold Dredging in Oregon - Originally published in 1939, this important publication on Oregon Mining has not been available for nearly seventy five years. Included are extremely rare insights into the history and day to day operations of the dragline and bucketline gold dredges that once worked the placer gold fields of South West and North East Oregon in decades gone by. Also included are details into the areas that were worked by gold dredges in Josephine, Jackson, Baker and Grant counties, as well as the economic factors that impacted this mining method. This volume also offers a unique look into the values of river bottom land in relation to both farming and mining, in how farm lands were mined, re-soiled and reclamated after the dredges worked them. Featured are hard to find maps of the gold dredge fields, as well as rare photographs from a bygone era. **8.5" X 11", 86 ppgs. Retail Price: $8.99**

Quick Silver Mining in Oregon - Originally published in 1963, this important publication on Oregon Mining has not been available for over fifty years. This publication includes details into the history and production of Elemental Mercury or Quicksilver in the State of Oregon. **8.5" X 11", 238 ppgs. Retail Price: $15.99**

Mines of the Greenhorn Mining District of Grant County Oregon - Originally published in 1948, this important publication on Oregon Mining has not been available for over sixty five years. In this publication are rare insights into the mines of the famous Greenhorn Mining District of Grant County, Oregon, especially the famous Morning Mine. Also included are details on the Tempest, Tiger, Bi-Metallic, Windsor, Psyche, Big Johnny, Snow Creek, Banzette and Paramount Mines, as well as prospects in the vicinities in the famous mining areas of Mormon Basin, Vinegar Basin and Desolation Creek. Included are hard to find mine maps and dozens of rare photographs from the bygone era of Grant County's rich mining history. **8.5" X 11", 72 ppgs. Retail Price: $9.99**

Geology of the Wallowa Mountains of Oregon: Part I (Volume 1) - Originally published in 1938, this important publication on Oregon Mining has not been available for nearly seventy five years. Included are details on the geology of this unique portion of North Eastern Oregon. This is the first part of a two book series on the area. Accompanying the text are rare photographs and historic maps. **8.5" X 11", 92 ppgs. Retail Price: $9.99**

Geology of the Wallowa Mountains of Oregon: Part II (Volume 2) - Originally published in 1938, this important publication on Oregon Mining has not been available for nearly seventy five years. Included are details on the geology of this unique portion of North Eastern Oregon. This is the first part of a two book series on the area. Accompanying the text are rare photographs and historic maps. **8.5" X 11", 94 ppgs. Retail Price: $9.99**

Field Identification of Minerals For Oregon Prospectors - Originally published in 1940, this important publication on Oregon Mining has not been available for nearly seventy five years. Included in this volume is an easy system for testing and identifying a wide range of minerals that might be found by prospectors, geologists and rockhounds in the State of Oregon, as well as in other locales. Topics include how to put together your own field testing kit and how to conduct rudimentary tests in the field. This volume is written in a clear and concise way to make it useful even for beginners. **8.5" X 11", 158 ppgs. Retail Price: $14.99**

The Bohemia Mining District of Oregon - Originally published in 1900, this important publication on Oregon Mining has not been available for over a century. Included in this volume are important insights into the famous Bohemia Mining District of Oregon, including the histories and locations of important gold mines in the area such as the Ophir Mine, Clarence, Acturas, Peek-a-boo, White Swan, Combination Mine, the Musick Mine, The California, White Ghost, The Mystery, Wall Street, Vesuvius, Story, Lizzie Bullock, Delta, Elsie Dora, Golden Slipper, Broadway, Champion Mine, Knott, Noonday, Helena, White Wings, Riverside and others. Also included are notes on the nearby Blue River Mining District. **8.5" X 11", 58 ppgs. Retail Price: $9.99**

The Gold Fields of Eastern Oregon - Unavailable since 1900, this publication was originally compiled by the Baker City Chamber of Commerce Offering important insights into the gold mining history of Eastern Oregon, "The Gold Fields of Eastern Oregon" sheds a rare light on many of the gold mines that were operating at the turn of the 19th Century in Baker County and Grant County in North Eastern Oregon. Some of the areas featured include the Cable Cove District, Baisely-Elhorn, Granite, Red Boy, Bonanza, Susanville, Sparta, Virtue, Vaughn, Sumpter, Burnt River, Rye Valley and other mining districts. Included is basic information on not only many gold mines that are well known to those interested in Eastern Oregon mining history, but also many mines and prospects which have been mostly lost to the passage of time. Accompanying are numerous rare photos **8.5" X 11", 78 ppgs. Retail Price: $10.99**

Gold Mining in Eastern Oregon - Originally published in 1938, this important publication on Oregon Mining has not been available for over a century. Included in this volume are important insights into the famous mining districts of Eastern Oregon during the late 1930's. Particular attention is given to those gold mines with milling and concentrating facilities in the Greenhorn, Red Boy, Alamo, Bonanza, Granite, Cable Cove, Cracker Creek, Virtue, Keating, Medical Springs, Sanger, Sparta, Chicken Creek, Mormon Basin, Connor Creek, Cornucopia and the Bull Run Mining Districts. Some of the mines featured include the Ben Harrison, North Pole-Columbia, Highland Maxwell, Baisley-Elkhorn, White Swan, Balm Creek, Twin Baby, Gem of Sparta, New Deal, Gleason, Gifford-Johnson, Cornucopia, Record, Bull Run, Orion and others. Of particular interest are the mill flow sheets and descriptions of milling operations of these mines. **8.5" X 11", 68 ppgs. Retail Price: $8.99**

The Gold Belt of the Blue Mountains of Oregon - Originally published in 1901, this important publication on Oregon Mining has not been available for over a century. Included in this volume are rare insights into the gold deposits of the Blue Mountains of North East Oregon, including the history of their early discovery and early production. Extensive details are offered on this important mining area's mineralogy and economic geology, as well as insights into nearby gold placers, silver deposits and copper deposits. Featured are the Elkhorn and Rock Creek mining districts, the Pocahontas district, Auburn and Minersville districts, Sumpter and Cracker Creek, Cable Cove, the Camp Carson district, Granite, Alamo, Greenhorn, Robinsonville, the Upper Burnt River Valley and Bonanza districts, Susanville, Quartzburg, Canyon Creek, Virtue, the Copper Butte district, the North Powder River, Sparta, Eagle Creek, Cornucopia, Pine Creek, Lower Powder River, the Upper Snake River Canyon, Rye Valley, Lower Burnt River Valley, Mormon Basin, the Malheur and Clarks Creek districts, Sutton Creek and others. Of particular interest are important details on numerous gold mines and prospects in these mining districts, including their locations, histories, geology and other important information, as well as information on silver, copper and fire opal deposits. **8.5" X 11", 250 ppgs. Retail Price: $24.99**

<u>Mining in the Cascades Range of Oregon</u> - Originally published in 1938, this important publication on Oregon Mining has not been available for over seventy five years. Included in this volume are rare insights into the gold mines and other types of metal mines in the Cascades Mountain Range of Oregon. Some of the important mining areas covered include the famous Bohemia Mining District, the North Santiam Mining District, Quartzville Mining District, Blue River Mining District, Fall Creek Mining District, Oakridge District, Zinc District, Buzzard-Al Sarena District, Grand Cove, Climax District and Barron Mining District. Of particular interest are important details on over 100 mines and prospects in these mining districts, including their locations, histories, geology and other important information. **8.5" X 11", 170 ppgs. Retail Price: $14.99**

<u>Beach Gold Placers of the Oregon Coast</u> - Originally published in 1934, this important publication on Oregon Mining has not been available for over 80 years. Included in this volume are rare insights into the beach gold deposits of the State of Oregon, including their locations, occurance, composition and geology. Of particular interest is information on placer platinum in Oregon's rich beach deposits. Also included are the locations and other information on some famous Oregon beach mines, including the Pioneer, Eagle, Chickamin, Iowa and beach placer mines north of the mouth of the Rogue River. **8.5" X 11", 60 ppgs. Retail Price: $8.99**

Idaho Mining Books

<u>Gold in Idaho</u> - Unavailable since the 1940's, this publication was originally compiled by the Idaho Bureau of Mines and includes details on gold mining in Idaho. Included is not only raw data on gold production in Idaho, but also valuable insight into where gold may be found in Idaho, as well as practical information on the gold bearing rocks and other geological features that will assist those looking for placer and lode gold in the State of Idaho. This volume also includes thirteen gold maps that greatly enhance the practical usability of the information contained in this small book detailing where to find gold in Idaho. **8.5" X 11", 72 ppgs. Retail Price: $9.99**

<u>Geology of the Couer D'Alene Mining District of Idaho</u> - Unavailable since 1961, this publication was originally compiled by the Idaho Bureau of Mines and Geology and includes details on the mining of gold, silver and other minerals in the famous Coeur D'Alene Mining District in Northern Idaho. Included are details on the early history of the Coeur D'Alene Mining District, local tectonic settings, ore deposit features, information on the mineral belts of the Osburn Fault, as well as detailed information on the famous Bunker Hill Mine, the Dayrock Mine, Galena Mine, Lucky Friday Mine and the infamous Sunshine Mine. This volume also includes sixteen hard to find maps. **8.5" X 11", 70 ppgs. Retail Price: $9.99**

<u>The Gold Camps and Silver Cities of Idaho</u> - Originally published in 1963, this important publication on Idaho Mining has not been available for nearly fifty years. Included are rare insights into the history of Idaho's Gold Rush, as well as the mad craze for silver in the Idaho Panhandle. Documented in fine detail are the early mining excitements at Boise Basin, at South Boise, in the Owyhees, at Deadwood, Long Valley, Stanley Basin and Robinson Bar, at Atlanta, on the famous Boise River, Volcano, Little Smokey, Banner, Boise Ridge, Hailey, Leesburg, Lemhi, Pearl, at South Mountain, Shoup and Ulysses, Yellow Jacket and Loon Creek. The story follows with the appearance of Chinese miners at the new mining camps on the Snake River, Black Pine, Yankee Fork, Bay Horse, Clayton, Heath, Seven Devils, Gibbonsville, Vienna and Sawtooth City. Also included are special sections on the Idaho Lead and Silver mines of the late 1800's, as well as the mining discoveries of the early 1900's that paved the way for Idaho's modern mining and mineral industry. Lavishly illustrated with rare historic photos, this volume provides a one of a kind documentary into Idaho's mining history that is sure to be enjoyed by not only modern miners and prospectors who still scour the hills in search of nature's treasures, but also those enjoy history and tromping through overgrown ghost towns and long abandoned mining camps. **8.5" X 11", 186 ppgs. Retail Price: $14.99**

<u>Ore Deposits and Mining in North Western Custer County Idaho</u> - Unavailable since 1913, this important publication was originally published by the Us Department of the Interior and has been unavailable for a century. Included are fine details on the geology, geography, gold placers and gold and silver bearing quartz veins of the mining region of North West Custer County, Idaho. Of particular interest is a rare look at the mines and prospects of the region, including those such as the Ramshorn Mine, SkyLark, Riverview, Excelsior, Beardsley, Pacific, Hoosier, Silver Brick, Forest Rose and dozens of others in the Bay Horse Mining District. Also covered are the mines of the Yankee Fork District such as the Lucky Boy, Badger, Black, Enterprise, Charles Dickens, Morrison, Golden Sunbeam, Montana, Golden Gate and others, as well as those in the Loon Mining District. **8.5" X 11", 126 ppgs. Retail Price: $12.99**

Gold Rush To Idaho - Unavailable since 1963, this important publication was originally published by the Idaho Bureau of Mines and has been unavailable for 50 years. "Gold Rush To Idaho" revisits the earliest years of the discovery of gold in Idaho Territory and introduces us to the conditions that the pioneer gold seekers met when they blazed a trail through the wilderness of Idaho's mountains and discovered the precious yellow metal at Oro Fino and Pierce. Subsequent rushes followed at places like Elk City, Newsome, Clearwater Station, Florence, Warrens and elsewhere. Of particular interest is a rare look at the hardships that the first miners in Idaho met with during their day to day existences and their attempts to bring law and order to their mining camps. **8.5" X 11", 88 ppgs. Retail Price: $9.99**

The Geology and Mines of Northern Idaho and North Western Montana - Unavailable since 1909, this important publication was originally published by the Us Department of the Interior and has been unavailable for a century. Included are fine details on the geology and geography of the mining regions of Northern Idaho and North Western Montana. Of particular interest is a rare look at the mines and prospects of the region, including those in the Pine Creek Mining District, Lake Pend Oreille district, Troy Mining District, Sylvanite District, Cabinet Mining District, Prospect Mining District and the Missoula Valley. Some of the mines featured include the Iron Mountain, Silver Butte, Snowshoe, Grouse Mountain Mine and others. **8.5" X 11", 142 ppgs. Retail Price: $12.99**

Mining in the Alturas Quadrangle of Blaine County Idaho - Unavailable since 1922, this important publication was originally published by the Idaho Bureau of Mines and has been unavailable for ninety years. Topics include the geology, rock formations and the formation of ore deposits in this important mining area of Idaho. Of particular focus is information on the local geology, quartz veins and ore deposits of this portion of Idaho. Included are hard to find details, including the descriptions and locations of numerous gold and silver mines in the area including the Silver King, Pilgrim, Columbia, Lone Jack, Sunbeam, Pride of the West, Lucky Boy, Scotia, Atlanta, Beaver-Bidwell and others mines and prospects. **8.5" X 11", 56 ppgs. Retail Price: $8.99**

Mining in Lemhi County Idaho - Originally published in 1913, this important book on Idaho Mining has not been available to miners for over a century. Included are rare insights into hundreds of gold, silver, copper and other mines in this famous Idaho mining area. Details include the locations, geology, history, production and other facts of the mines of this region, not only gold and silver hardrock mines, but also gold placer mines, lead-silver deposits, copper mines, cobalt-nickel deposits, tungsten and tin mines. It is lavishly illustrated with hard to find photos of the period and rare mining maps. Some of the vicinities featured include the Nicholia Mining District, Spring Mountain District, Texas District, Blue Wing District, Junction District, McDevitt District, Pratt Creek, Eldorado District, Kirtley Creek, Carmen Creek, Gibbonsville, Indian Creek, Mineral Hill District, Mackinaw, Eureka District, Blackbird District, YellowJacket District, Gravel Range District, Junction District, Parker Mountain and other mining districts. **8.5" X 11", 226 ppgs. Retail Price: $19.99**

Utah Mining Books

Fluorite in Utah - Unavailable since 1954, this publication was originally compiled by the USGS, State of Utah and U.S. Atomic Energy Commission and details the mining of fluorspar, also known as fluorite in the State of Utah. Included are details on the geology and history of fluorspar (fluorite) mining in Utah, including details on where this unique gem mineral may be found in the State of Utah. **8.5" X 11", 60 ppgs. Retail Price: $8.99**

California Mining Books

The Tertiary Gravels of the Sierra Nevada of California - Mining historian Kerby Jackson introduces us to a classic mining work by Waldemar Lindgren in this important re-issue of The Tertiary Gravels of the Sierra Nevada of California. Unavailable since 1911, this publication includes details on the gold bearing ancient river channels of the famous Sierra Nevada region of California. **8.5" X 11", 282 ppgs. Retail Price: $19.99**

The Mother Lode Mining Region of California - Unavailable since 1900, this publication includes details on the gold mines of California's famous Mother Lode gold mining area. Included are details on the geology, history and important gold mines of the region, as well as insights into historic mining methods, mine timbering, mining machinery, mining bell signals and other details on how these mines operated. Also included are insights into the gold mines of the California Mother Lode that were in operation during the first sixty years of California's mining history. **8.5" X 11", 176 ppgs. Retail Price: $14.99**

Lode Gold of the Klamath Mountains of Northern California and South West Oregon - Unavailable since 1971, this publication was originally compiled by Preston E. Hotz and includes details on the lode mining districts of Oregon and California's Klamath Mountains. Included are details on the geology, history and important lode mines of the French Gulch, Deadwood, Whiskeytown, Shasta, Redding, Muletown, South Fork, Old Diggings, Dog Creek (Delta), Bully Choop (Indian Creek), Harrison Gulch, Hayfork, Minersville, Trinity Center, Canyon Creek, East Fork, New River, Denny, Liberty (Black Bear), Cecilville, Callahan, Yreka, Fort Jones and Happy Camp mining districts in California, as well as the Ashland, Rogue River, Applegate, Illinois River, Takilma, Greenback, Galice, Silver Peak, Myrtle Creek and Mule Creek districts of South Western Oregon. Also included are insights into the mineralization and other characteristics of this important mining region. **8.5" X 11", 100 ppgs. Retail Price: $10.99**

Mines and Mineral Resources of Shasta County, Siskiyou County, Trinity County: California - Unavailable since 1915, this publication was originally compiled by the California State Mining Bureau and includes details on the gold mines of this area of Northern California. Also included are insights into the mineralization and other characteristics of this important mining region, as well as the location of historic gold mines. **8.5″ X 11″, 204 ppgs. Retail Price: $19.99**

Geology of the Yreka Quadrangle, Siskiyou County, California - Unavailable since 1977, this publication was originally compiled by Preston E. Hotz and includes details on the geology of the Yreka Quadrangle of Siskiyou County, California. Also included are insights into the mineralization and other characteristics of this important mining region. **8.5″ X 11″, 78 ppgs. Retail Price: $7.99**

Mines of San Diego and Imperial Counties, California - Originally published in 1914, this important publication on California Mining has not been available for a century. This publication includes important information on the early gold mines of San Diego and Imperial County, which were some of the first gold fields mined in California by early Spanish and Mexican miners before the 49ers came on the scene. Included are not only details on early mining methods in the area, production statistics and geological information, but also the location of the early gold mines that make California "The Golden State". Also included are details on the mining of other minerals such as silver, lead, zinc, manganese, tungsten, vanadium, asbestos, barite, borax, cement, clay, dolomite, fluospar, gem stones, graphite, marble, salines, petroleum, stronium, talc and others. **8.5″ X 11″, 116 ppgs. Retail Price: $12.99**

Mines of Sierra County, California - Unavailable since 1920, this publication was originally compiled by the California State Mining Bureau and includes details on the gold mines of Sierra County, California. Also included are insights into the mineralization and other characteristics of this important mining region, as well as the location of historic gold mines. **8.5″ X 11″, 156 ppgs. Retail Price: $19.99**

Mines of Plumas County, California - Unavailable since 1918, this publication was originally compiled by the California State Mining Bureau and includes details on the gold mines of Plumas County, California. Also included are insights into the mineralization and other characteristics of this important mining region, as well as the location of historic gold mines. **8.5″ X 11″, 200 ppgs. Retail Price: $19.99**

Mines of El Dorado, Placer, Sacramento and Yuba Counties, California - Originally published in 1917, this important publication on California Mining has not been available for nearly a century. This publication includes important information on the early gold mines of El Dorado County, Placer County, Sacramento County and Yuba County, which were some of the first gold fields mined by the Forty-Niners during the California Gold Rush. Included are not only details on early mining methods in the area, production statistics and geological information, but also the location of the early gold mines that helped make California "The Golden State". Also included are insights into the early mining of chrome, copper and other minerals in this important mining area. **8.5″ X 11″, 204 ppgs. Retail Price: $19.99**

Mines of Los Angeles, Orange and Riverside Counties, California - Originally published in 1917, this important publication on California Mining has not been available for nearly a century. This publication includes important information on the early gold mines of Los Angeles County, Orange County and Riverside County, which were some of the first gold fields mined in California by early Spanish and Mexican miners before the 49ers came on the scene. Included are not only details on early mining methods in the area, production statistics and geological information, but also the location of the early gold mines that helped make California "The Golden State". **8.5″ X 11″, 146 ppgs. Retail Price: $12.99**

Mines of San Bernadino and Tulare Counties, California - Originally published in 1917, this important publication on California Mining has not been available for nearly a century. This publication includes important information on the early gold mines of San Bernadino and Tulare County, which were some of the first gold fields mined in California by early Spanish and Mexican miners before the 49ers came on the scene. Included are not only details on early mining methods in the area, production statistics and geological information, but also the location of the early gold mines that helped make California "The Golden State". Also included are details on the mining of other minerals such as copper, iron, lead, zinc, manganese, tungsten, vanadium, asbestos, barite, borax, cement, clay, dolomite, fluospar, gem stones, graphite, marble, salines, petroleum, stronium, talc and others. **8.5″ X 11″, 200 ppgs. Retail Price: $19.99**

Chromite Mining in The Klamath Mountains of California and Oregon - Unavailable since 1919, this publication was originally compiled by J.S. Diller of the United States Department of Geological Survey and includes details on the chromite mines of this area of Northern California and Southern Oregon. Also included are insights into the mineralization and other characteristics of this important mining region, as well as the location of historic mines. Also included are insights into chromite mining in Eastern Oregon and Montana. **8.5″ X 11″, 98 ppgs. Retail Price: $9.99**

Mines and Mining in Amador, Calaveras and Tuolumne Counties, California - Unavailable since 1915, this publication was originally compiled by William Tucker and includes details on the mines and mineral resources of this important California mining area. Included are details on the geology, history and important gold mines of the region, as well as insights into other local mineral resources such as asbestos, clay, copper, talc, limestone and others. Also included are insights into the mineralization and other characteristics of this important portion of California's Mother Lode mining region. **8.5" X 11", 198 ppgs. Retail Price: $14.99**

The Cerro Gordo Mining District of Inyo County California - Unavailable since 1963, this publication was originally compiled by the United States Department of Interior. Included are insights into the mineralization and other characteristics of this important mining region of Southern California. Topics include the mining of gold and silver in this important mining district in Inyo County, California, including details on the history, production and locations of the Cerro Gordo Mine, the Morning Star Mine, Estelle Tunnel, Charles Lease Tunnel, Ignacio, Hart, Crosscut Tunnel, Sunset, Upper Newtown, Newtown, Ella, Perseverance, Newsboy, Belmont and other silver and gold mines in the Cerro Gordo Mining District. This volume also includes important insights into the fossil record, geologic formations, faults and other aspects of economic geology in this California mining district. **8.5" X 11", 104 ppgs. Retail Price: $10.99**

Mining in Butte, Lassen, Modoc, Sutter and Tehama Counties of California - Unavailable since 1917, this publication was originally compiled by the United States Department of Interior. Included are insights into the mineralization and other characteristics of this important mining region of California. Topics include the mining of asbestos, chromite, gold, diamonds and manganese in Butte County, the mining of gold and copper in the Hayden Hill and Diamond Mountain mining districts of Lassen County, the mining of coal, salt, copper and gold in the High Grade and Winters mining districts of Modoc County, gold mining in Sutter County and the mining of gold, chromite, manganese and copper in Tehama County. This volume also includes the production records and locations of numerous mines in this important mining region. **8.5" X 11", 114 ppgs. Retail Price: $11.99**

Mines of Trinity County California - Originally published in 1965, this important publication on California Mining has not been available for nearly fifty years. This publication includes important information on mines and mining in Trinity County, California, as well insights into the mineralization and geology of this important mining area in Northern California. Included are extensive details on hardrock and placer gold mines and prospects, including charts showing the locations of these historic mines.. **8.5" X 11", 144 ppgs. Retail Price: $12.99**

Mines of Kern County California - Originally published in 1962, this important publication on California Mining has not been available for nearly fifty years. This publication includes important information on mines and mining in Kern County, California, as well insights into the mineralization and geology of this important mining area in California. Included are extensive details on hardrock and placer gold mines and prospects, including charts showing the locations of these historic mines. **8.5" X 11", 398 ppgs. Retail Price: $24.99**

Mines of Calaveras County California - Originally published in 1962, this important publication on California Mining has not been available for nearly fifty years. This publication includes important information on mines and mining in Calaveras County, California, as well insights into the mineralization and geology of this important mining area in Northern California. Included are extensive details on hardrock and placer gold mines and prospects, including charts showing the locations of these historic mines. **8.5" X 11", 236 ppgs. Retail Price: $19.99**

Lode Gold Mining in Grass Valley California - Unavailable since 1940, this publication was originally compiled by the United States Department of Interior. Included are insights into the gold mineralization and other characteristics of this important mining region of Nevada County, California. This volume also includes important insights into the geologic formations, faults and other aspects of economic geology in this California mining district. Of particular interest are the fine details on many hardrock gold mines in the area, including their locations, histories, development and mineralization. Some of the mines featured include the Gold Hill Mine, Massachusetts Hill, Boundary, Peabody, Golden Center, North Star, Omaha, Lone Jack, Homeward Bound, Hartery, Wisconsin, Allison Ranch, Phoenix, Kate Hayes, W.Y.O.D., Empire, Rich Hill, Daisy Hill, Orleans, Sultana, Centennial, Conlin, Ben Franklin, Crown Point and many others. **8.5" X 11", 148 ppgs. Retail Price: $12.99**

Lode Mining in the Alleghany District of Sierra County California - Unavailable since 1913, this publication was originally compiled by the United States Department of Interior. Included are insights into the mineralization and other characteristics of this important mining region of Sierra County. Included are details on the history, production and locations of numerous hardrock gold mines in this famous California area, including the Tightner Mine, Minnie D., Osceola, Eldorado, Twenty One, Sherman, Kenton, Oriental, Rainbow, Plumbago, Irelan, Gold Canyon, North Fork, Federal, Kate Hardy and others. This volume also includes important insights into the fossil record, geologic formations, faults and other aspects of economic geology in this California mining district. **8.5" X 11", 48 ppgs. Retail Price: $7.99**

Six Months In The Gold Mines During The California Gold Rush - Unavailable since 1850, this important work is a first hand account of one "49'ers" personal experience during the great California Gold Rush, shedding important light on one of the most exciting periods in the history of not only California, but also the world. Compiled from journals written between 1847 and 1849 by E. Gould Buffum, a native of New York, "Six Months In The Gold Mines During The California Gold Rush" offers a rare look into the day to day lives of the people who came to California to work in her gold mines when the state was still a great frontier. **8.5" X 11", 290 ppgs. Retail Price: $19.99**

Quartz Mines of the Grass Valley Mining District of California - Unavailable since 1867, this important publication has not been available since those days. This rare publication offers a short dissertation on the early hardrock mines in this important mining district in the California Mother Lode region between the 1850's and 1860's. Also included are hard to find details on the mineralization and locations of these mines, as well as how they were operated in those day. **8.5" X 11", 44 ppgs. Retail Price: $8.99**

Alaska Mining Books

Ore Deposits of the Willow Creek Mining District, Alaska - Unavailable since 1954, this hard to find publication includes valuable insights into the Willow Creek Mining District near Hatcher Pass in Alaska. The publication includes insights into the history, geology and locations of the well known mines in the area, including the Gold Cord, Independence, Fern, Mabel, Lonesome, Snowbird, Schroff-O'Neil, High Grade, Marion Twin, Thorpe, Webfoot, Kelly-Willow, Lane, Holland and others. **8.5" X 11", 96 ppgs. Retail Price: $9.99**

The Juneau Gold Belt of Alaska - Unavailable since 1906, this hard to find publication includes valuable insights into the gold mines around Juneau, Alaska. The publication includes important details into the history, geology and locations of the well known gold mines and prospects in the area, including those around Windham Bay, Holkham Bay, Port Snettisham, on Grindstone and Rhine Creeks, Gold Creek, Douglas Island, Salmon Creek, Lemon Creek, Nugget Creek, from the Mendenhall River to Berners Bay, McGinnis Creek, Montana Creek, Peterson Creek, Windfall Creek, the Eagle River, Yankee Basin, Yankee Curve, Kowee Creek and elsewhere. Not only are gold placer mines included, but also hardrock gold mines. **8.5" X 11", 224 ppgs. Retail Price: $19.99**

Arizona Mining Books

Mines and Mining in Northern Yuma County Arizona - Originally published in 1911, this important publication on Arizona Mining has not been available for over a hundred years. Included are rare insights into the gold, silver, copper and quicksilver mines of Yuma County, Arizona together with hard to find maps and photographs. Some of the mines and mining districts featured include the Planet Copper Mine, Mineral Hill, the Clara Consolidated Mine, Viati Mine, Copper Basin prospect, Bowman Mine, Quartz King, Billy Mack, Carnation, the Wardwell and Osbourne, Valensuella Copper, the Mariquita, Colonial Mine, the French American, the New York-Plomosa, Guadalupe, Lead Camp, Mudersbach Copper Camp, Yellow Bird, the Arizona Northern (Salome Strike), Bonanza (Harqua Hala), Golden Eagle, Hercules, Socorro and others. **8.5" X 11", 144 ppgs. Retail Price: $11.99**

The Aravaipa and Stanley Mining Districts of Graham County Arizona - Originally published in 1925, this important publication on Arizona Mining has not been available for nearly ninety years. Included are rare insights into the gold and silver mines of these two important mining districts, together with hard to find maps. **8.5" X 11", 140 ppgs. Retail Price: $11.99**

Gold in the Gold Basin and Lost Basin Mining Districts of Mohave County, Arizona - This volume contains rare insights into the geology and gold mineralization of the Gold Basin and Lost Basin Mining Districts of Mohave County, Arizona that will be of benefit to miners and prospectors. Also included is a significant body of information on the gold mines and prospects of this portion of Arizona. This volume is lavishly illustrated with rare photos and mining maps. **8.5" X 11", 188 ppgs. Retail Price: $19.99**

Mines of the Jerome and Bradshaw Mountains of Arizona - This important publication on Arizona Mining has not been available for ninety years. This volume contains rare insights into the geology and ore deposits of the Jerome and Bradshaw Mountains of Arizona that will be of benefit to miners and prospectors who work those areas. Included is a significant body of information on the mines and prospects of the Verde, Black Hills, Cherry Creek, Prescott, Walker, Groom Creek, Hassayampa, Bigbug, Turkey Creek, Agua Fria, Black Canyon, Peck, Tiger, Pine Grove, Bradshaw, Tintop, Humbug and Castle Creek Mining Districts. This volume is lavishly illustrated with rare photos and mining maps. **8.5" X 11", 218 ppgs. Retail Price: $19.99**

The Ajo Mining District of Pima County Arizona - This important publication on Arizona Mining has not been available for nearly seventy years. This volume contains rare insights into the geology and mineralization of the Ajo Mining District in Pima County, Arizona and in particular the famous New Cornelia Mine. **8.5" X 11", 126 ppgs. Retail Price: $11.99**

Mining in the Santa Rita and Patagonia Mountains of Arizona - Originally published in 1915, this important publication on Arizona Mining has not been available for nearly a century. Included are rare insights into hundreds of gold, silver, copper and other mines in this famous Arizona mining area. Details include the locations, geology, history, production and other facts of the mines of this region. **8.5" X 11", 394 ppgs. Retail Price: $24.99**

Mining in the Bisbee Quadrangle of Arizona - Originally published in 1906, this important publication on Arizona Mining has not been available for nearly a century. Included are rare insights into hundreds of gold, silver, copper and other mines in this famous Arizona mining area. Details include the locations, geology, history, production and other facts of the mines of this important mining region. **8.5" X 11", 188 ppgs. Retail Price: $14.99**

Montana Mining Books

A History of Butte Montana: The World's Greatest Mining Camp - First published in 1900 by H.C. Freeman, this important publication sheds a bright light on one of the most important mining areas in the history of The West. Together with his insights, as well as rare photographs of the periods, Harry Freeman describes Butte and its vicinity from its early beginnings, right up to its flush years when copper flowed from its mines like a river. At the time of publication, Butte, Montana was known worldwide as "The Richest Mining Spot On Earth" and produced not only vast amounts of copper, but also silver, gold and other metals from its mines. Freeman illustrates, with great detail, the most important mines in the vicinity of Butte, providing rare details on their owners, their history and most importantly, how the mines operated and how their treasures were extracted. Of particular interest are the dozens of rare photographs that depict mines such as the famous Anaconda, the Silver Bow, the Smoke House, Moose, Paulin, Buffalo, Little Minah, the Mountain Consolidated, West Greyrock, Cora, the Green Mountain, Diamond, Bell, Parnell, the Neversweat, Nipper, Original and many others. **8.5" X 11", 142 ppgs. Retail Price: $12.99**

The Butte Mining District of Montana - This important publication on Montana Mining has not been available for over a century. Included are rare insights into the gold, copper and silver mines of Butte, Montana together with hard to find maps and photographs. Some of the topics include the early history of gold, silver and copper mining in the Butte area, insight into the geology of its mining areas, the local distribution of gold, silver and copper ores, as well their composition and how to identify them. Also included are detailed facts about the mines in the Butte Mining District, including the famous Anaconda Mine, Gagnon, Parrot, Blue Vein, Moscow, Poulin, Stella, Buffalo, Green Mountain, Wake Up Jim, the Diamond-Bell Group, Mountain Consolidated, East Greyrock, West Greyrock, Snowball, Corra, Speculator, Adirondack, Miners Union, the Jessie-Edith May Group, Otisco, Iduna, Colorado, Lizzie, Cambers, Anderson, Hesperus, Preferencia and dozens of others. **8.5" X 11", 298 ppgs. Retail Price: $24.99**

Mines of the Helena Mining Region of Montana - This important publication on Montana Mining has not been available for over a century. Included are rare insights into the gold, copper and silver mines of the vicinity of Helena, Montana, including the Marysville Mining District, Elliston Mining District, Rimini Mining District, Helena Mining District, Clancy Mining District, Wickes Mining District, Boulder and Basin Mining Districts and the Elkhorn Mining District. Some of the topics include the early history of gold, silver and copper mining in the Helena area, insight into the geology of its mining areas, the local distribution of gold, silver and copper ores, as well their composition and how to identify them. Also included are detailed facts, history, geology and locations of over one hundred gold, silver and copper mines in the area . **8.5" X 11", 162 ppgs, Retail Price: $14.99**

Mines and Geology of the Garnet Range of Montana - This important publication on Montana Mining has not been available for over a century. Included are rare insights into the gold, copper and silver mines of the vicinity of this important mining area of Montana. Some of the topics include the early history of gold, silver and copper mining in the Garnet Mountains, insight into the geology of its mining areas, the local distribution of gold, silver and copper ores, as well their composition and how to identify them. Also included are detailed facts, history, geology and locations of numerous gold, silver and copper mines in the area . **8.5" X 11", 100 ppgs, Retail Price: $11.99**

Mines and Geology of the Philipsburg Quadrangle of Montana - This important publication on Montana Mining has not been available for over a century. Included are rare insights into the gold, copper and silver mines of the vicinity of this important mining area of Montana. Some of the topics include the early history of gold, silver and copper mining in the Philipsburg Quadrangle, insight into the geology of its mining areas, the local distribution of gold, silver and copper ores, as well their composition and how to identify them. Also included are detailed facts, history, geology and locations of over one hundred gold, silver and copper mines in the area **8.5" X 11", 290 ppgs, Retail Price: $24.99**

Geology of the Marysville Mining District of Montana - Included are rare insights into the mining geology of the Marysville Mining District. Some of the topics include the early history of gold, silver and copper mining in the area, insight into the geology of its mining areas, the local distribution of gold, silver and copper ores, as well their composition and how to identify them. Also included are detailed facts, history, geology and locations of gold, silver and copper mines in the area **8.5" X 11", 198 ppgs, Retail Price: $19.99**

<u>**The Geology and Mines of Northern Idaho and North Western Montana**</u>

See listing under Idaho.

Nevada Mining Books

<u>**The Bull Frog Mining District of Nevada**</u> - Unavailable since 1910, this publication was originally compiled by the United States Department of Interior. This volume also includes important insights into the geologic formations, faults and other aspects of economic geology in this Nevada mining district. Of particular interest are the fine details on many mines in the area, including their locations, histories, development and mineralization. Some of the mines featured include the National Bank Mine, Providence, Gibraltor, Tramps, Denver, Original Bullfrog, Gold Bar, Mayflower, Homestake-King and other mines and prospects. **8.5" X 11", 152 ppgs, Retail Price: $14.99**

<u>History of the Comstock Lode</u> - Unavailable since 1876, this publication was originally released by John Wiley & Sons. This volume also includes important insights into the famous Comstock Lode of Nevada that represented the first major silver discovery in the United States. During its spectacular run, the Comstock produced over 192 million ounces of silver and 8.2 million ounces of gold. Not only did the Comstock result in one of the largest mining rushes in history and yield immense fortunes for its owners, but it made important contributions to the development of the State of Nevada, as well as neighboring California. Included here are important details on not only the early development and history of the Comstock, but also rare early insight into its mines, ore and its geology.**8.5" X 11", 244 ppgs, Retail Price: $19.99**

Colorado Mining Books

<u>**Ores of The Leadville Mining District**</u> - Unavailable since 1926, this publication was originally compiled by the United States Department of Interior. This volume also includes important insights into the ores and mineralization of the Leadville Mining District in Colorado. Topics include historic ore prospecting methods, local geology, insights into ore veins and stockworks, the local trend and distribution of ore channels, reverse faults, shattered rock above replacement ore bodies, mineral enrichment in oxidized and sulphide zones and more. **8.5" X 11", 66 ppgs, Retail Price: $8.99**

<u>**Mining in Colorado**</u> - Unavailable since 1926, this publication was originally compiled by the United States Department of Interior. This volume also includes important insights into the mining history of Colorado from its early beginnings in the 1850's right up to the mid 1920's. Not only is Colorado's gold mining heritage included, but also its silver, copper, lead and zinc mining industry. Each mining area is treated separately, detailing the development of Colorado's mines on a county by county basis. **8.5" X 11", 284 ppgs, Retail Price: $19.99**

<u>Gold Mining in Gilpin County Colorado</u> - Unavailable since 1876, this publication was originally compiled by the Register Steam Printing House of Central City, Colorado. A rare glimpse at the gold mining history and early mines of Gilpin County, Colorado from their first discovery in the 1850's up to the "flush years" of the mid 1870's. Of particular interest is the history of the discovery of gold in Gilpin County and details about the men who made those first strikes. Special focus is given to the early gold mines and first mining districts of the area, many of which are not detailed in other books on Colorado's gold mining history. **8.5" X 11", 156 ppgs, Retail Price: $12.99**

<u>Mining in the Gold Brick Mining District of Colorado</u> - Important insights into the history of the Gold Brick Mining District, as well as its local geography and economic geology. Also included are the histories and locations of historic mines in this important Colorado Mining District, including the Cortland, Carter, Raymond, Gold Links, Sacramento, Bassick, Sandy Hook, Chronicle, Grand Prize, Chloride, Granite Mountain, Lucille, Gray Mountain, Hilltop, Maggie Mitchell, Silver Islet, Revenue, Roosevelt, Carbonate King and others. In addition to hardrock mining, are also included are details on gold placer mining in this portion of Colorado. **8.5" X 11", 140 ppgs, Retail Price: $12.99**

Washington Mining Books

<u>**The Republic Mining District of Washington**</u> - Unavailable since 1910, this important publication was originally published by the Washington Geologic Survey and has been unavailable for a century. Topics include the geology, rock formations and the formation of ore deposits in this important mining area of Washington State. Also included are hard to find details on the geology, history and locations of dozens of mines in the area. Some of the mines featured include the New Republic Mine, Ben Hur, Morning Glory, the South Republic Mine, Quilp, Surprise, Black Tail, Lone Pine, San Poil, Mountain Lion, Tom Thumb, Elcaliph and many others. **8.5" X 11", 94 ppgs, Retail Price: $10.99**

The Myers Creek and Nighthawk Mining Districts of Washington - Unavailable since 1911, this important publication was originally published by the Washington Geologic Survey and has been unavailable for a century. Topics include the geology, rock formations and the formation of ore deposits in these important mining areas of Washington State. Also included are hard to find details on the geology, history and locations of dozens of mines in the area. Some of the mines featured include the Grant Mine, Monterey, Nip and Tuck, Myers Creek, Number Nine, Neutral, Rainbow, Aztec, Crystal Butte, Apex, Butcher Boy, Molson, Mad River, Olentangy, Delate, Kelsey, Golden Chariot, Okanogan, Ohio, Forty-Ninth Parallel, Nighthawk, Favorite, Little Chopaka, Summit, Number One, California, Peerless, Caaba, Prize Group, Ruby, Mountain Sheep, Golden Zone, Rich Bar, Similkameen, Kimberly, Triune, Hiawatha, Trinity, Hornsilver, Maquae, Bellevue, Bullfrog, Palmer Lake, Ivanhoe, Copper World and many others.
 8.5" X 11", 136 ppgs, Retail Price: $12.99

The Blewett Mining District of Washington - Unavailable since 1911, this important publication was originally published by the Washington Geologic Survey and has been unavailable for a century. Topics include the geology, rock formations and the formation of ore deposits in this important mining area of Washington State. Also included are hard to find details on the geology, history and locations of dozens of mines in the area. Some of the mines featured include the Washington Meteor, Alta Vista, Pole Pick, Blinn, North Star, Golden Eagle, Tip Top, Wilder, Golden Guinea, Lucky Queen, Blue Bell, Prospect, Homestake, Lone Rock, Johnson, and others. **8.5" X 11", 134 ppgs, Retail Price: $12.99**

Silver Mining In Washington - Unavailable since 1955, this important publication was originally published by the Washington Geologic Survey. Featured are the hard to find locations and details pertaining to Washington's silver mines. **8.5" X 11", 180 ppgs, Retail Price: $15.99**

The Mines of Snohomish County Washington - Unavailable since 1942, this important publication was originally published by the Washington Geologic Survey and has been unavailable for seventy years. Featured are details on a large number of gold, silver, copper, lead and other metallic mineral mines. Included are the locations of each historic mine, along with information on the commodity produced. **8.5" X 11", 98 ppgs, Retail Price: $10.99**

The Mines of Chelan County Washington - Unavailable since 1943, this important publication was originally published by the Washington Geologic Survey and has been unavailable for seventy years. Featured are details on a large number of gold, silver, copper, lead and other metallic mineral mines. Included are the locations of each historic mine, along with information on the commodity. **8.5" X 11", 88 ppgs, Retail Price: $9.99**

Metal Mines of Washington - Unavailable since 1921, this important publication was originally published by the Washington Geologic Survey and has been unavailable for nearly ninety years. Widely considered a masterpiece on the Washington Mining Industry, "Metal Mines of Washington" sheds light on the important details of Washington's early mining years. Featured are details on hundreds of gold, silver, copper, lead and other metallic mineral mines. Included are hard to find details on the mineral resources of this state, as well as the locations of historic mines. Lavishly illustrated with maps and historic photos and complete with a glossary to explain any technical terms found in the text, this is one of the most important works on mining in the State of Washington. No prospector or miner should be without it if they are interested in mining in Washington. **8.5" X 11", 396 ppgs, Retail Price: $24.99**

Gem Stones In Washington - Unavailable since 1949, this important publication was originally published by the Washington Geologic Survey and has been unavailable since first published. Included are details on where to find naturally occurring gem stones in the State of Washington, including quartz crystal, amethyst, smoky quartz, milky quartz, agates, bloodstone, carnelian, chert, flint, jasper, onyx, petrified wood, opal, fire opal, hyalite and others. **8.5" X 11", 54 ppgs, Retail Price: $8.99**

The Covada Mining District of Washington - Unavailable since 1913, this important publication was originally published by the Washington Geologic Survey and has been unavailable for a century. Topics include the geology, rock formations and the formation of ore deposits in this important mining area of Washington State. Also included are hard to find details on the geology, history and locations of dozens of mines in the area. Some of the mines featured include the Admiral, Advance, Algonkian, Big Bug, Big Chief, Big Joker, Black Hawk, Black Tail, Black Thorn, Captain, Cherokee Strip, Colorado, Dan Patch, Dead Shot, Etta, Good Ore, Greasy Run, Great Scott, Idora, IXL, Jay Bird, Kentucky Bell, King Solomon, Laurel, Laura S, Little Jay, Meteor, Neglected, Northern Light, Old Nell, Plymouth Rock, Polaris, Quandary, Reserve, Shoo Fly, Silver Plume, Three Pines, Vernie, White Rose and dozens of others. **8.5" X 11", 114 ppgs, Retail Price: $10.99**

The Index Mining District of Washington - Unavailable since 1912, this important publication was originally published by the Washington Geologic Survey and has been unavailable for a century. Topics include the geology, rock formations and the formation of ore deposits in this important mining area of Washington State. Also included are hard to find details on the geology, history and locations of dozens of mines in the area. Some of the mines featured include the Sunset, Non-Pareil, Ethel Consolidated, Kittaning, Merchant, Homestead, Co-operative, Lost Creek, Uncle Sam, Calumet, Florence-Rae, Bitter Creek, Index Peacock, Gunn Peak, Helena, North Star, Buckeye. Copper Bell, Red Cross and others. **8.5" X 11", 114 ppgs, Retail Price: $11.99**

Mining & Mineral Resources of Stevens County Washington - Unavailable since 1920, this important publication was originally published by the Washington Geologic Survey and has been unavailable for a century. Topics include the geology, rock formations and the formation of ore deposits in these important mining areas of Washington State. Also included are hard to find details on the geology, history and locations of hundreds of mines in the area. 8.5" X 11", 372 ppgs, Retail Price: $24.99

The Mines and Geology of the Loomis Quadrangle Okanogan County, Washington - Unavailable since 1972, this important publication was originally published by the Washington Geologic Survey and has been unavailable for a century. Topics include the geology, rock formations and the formation of ore deposits in this important mining area of Washington State. Also included are hard to find details on the geology, history and locations of dozens of gold, copper, silver and other mines in the area. 8.5" X 11", 150 ppgs, Retail Price: $12.99

The Conconully Mining District of Okanogan County Washington - Unavailable since 1973, this important publication was originally published by the Washington Geologic Survey and has been unavailable for a century. Topics include the geology, rock formations and the formation of ore deposits in this important mining area of Washington State, which also includes Salmon Creek, Blue Lake and Galena. Also included are hard to find details on the geology, mining history and locations of dozens of mines in the area. Some of the mines include Arlington, Fourth of July, Sonny Boy, First Thought, Last Chance, War Eagle-Peacock, Wheeler, Mohawk, Lone Star, Woo Loo Moo Loo, Keystone, Hughes, Plant-Callahan, Johnny Boy, Leuena, Gubser, John Arthur, Tough Nut, Homestake, Key and many others 8.5" X 11", 68 ppgs, Retail Price: $8.99

Wyoming Mining Books

Mining in the Laramie Basin of Wyoming - Unavailable since 1909, this publication was originally compiled by the United States Department of Interior. Also included are insights into the mineralization and other characteristics of this important mining region, especially in regards to coal, limestone, gypsum, bentonite clay, cement, sand, clay and copper. 8.5" X 11", 104 ppgs, Retail Price: $11.99

New Mexico Mining Books

The Mogollon Mining District of New Mexico - Unavailable since 1927, this important publication was originally published by the US Department of Interior and has been unavailable for 80 years. Topics include the geology, rock formations and the formation of ore deposits in this important mining area in New Mexico. Of particular focus is information on the history and production of the ore deposits in this area, their form and structure, vein filling, their paragenesis, origins and ore shoots, as well as oxidation and supergene enrichment. Also included are hard to find details, including the descriptions and locations of numerous gold, silver and other types of mines, including the Eureka, Pacific, South Alpine, Great Western, Enterprise, Buffalo, Mountain View, Floride, Gold Dust, Last Chance, Deadwood, Confidence, Maud S., Deep Down, Little Fanney, Trilby, Johnson, Alberta, Comet, Golden Eagle, Cooney, Queen, the Iron Crown, Eberle, Clifton, Andrew Jackson mine, Mascot and others. 8.5" X 11", 144 ppgs, Retail Price: $12.99

The Percha Mining District of Kingston New Mexico - Unavailable since 1883, this important publication was originally published by the Kingston Tribune and has been unavailable for over one hundred and thirty five years. Having been written during the earliest years of gold and silver mining in the Percha Mining District, unlike other books on the subject, this work offers the unique perspective of having actually been written while the early mining history of this area was still being made. In fact, the work was written so early in the development of this area that many of the notable mines in the Percha District were less than a few years old and were still being operated by their original discoverers with the same enthusiasm as when they were first located. Included are hard to find details on the very earliest gold and silver mines of this important mining district near Kingston in Sierra County, New Mexico. 8.5" X 11", 68 ppgs, Retail Price: $9.99

East Coast Mining Books

The Gold Fields of the Southern Appalachians - Unavailable since 1895, this important publication was originally published by the US Department of Interior and has been unavailable for nearly 120 years. Topics include the geology, rock formations and the formation of ore deposits in this important mining area of the American South. Of particular focus is information on the history and statistics of the ore deposits in this area, their form and structure and veins. Also included are details on the placer gold deposits of the region. The gold fields of the Georgian Belt, Carolinian Belt and the South Mountain Mining District of North Carolina are all treated in descriptive detail. Included are hard to find details, including the descriptions and locations of numerous gold mines in Georgia, North Carolina and elsewhere in the American South. Also included are details on the gold belts of the British Maritime Provinces and the Green Mountains. 8.5" X 11", 104 ppgs, Retail Price: $9.99

Gold Rush Tales Series

Millions in Siskiyou County Gold - In this first volume of the "Gold Rush Tales" series, leading mining historian and editor Kerby Jackson, introduces us to the story of how millions of dollars worth of gold was discovered in Siskiyou County during the California Gold Rush. Lavishly illustrated with photos from the 19th Century, this hard to find information was first published in 1897 and sheds important light onto the gold rush era in Siskiyou County, California and the experiences of the men who dug for the gold and actually found it. **8.5" X 11", 82 ppgs, Retail Price: $9.99**

The California Rand in the Days of '49 - In this second volume of the "Gold Rush Tales" series, leading mining historian and editor Kerby Jackson, introduces us to four tales from the California Gold Rush. Lavishly illustrated with photos from the 19th Century, this hard to find information was first published in 1890's and includes the stories of "California's Rand", details about Chinese miners, how one early miner named Baker struck it rich and also the story of Alphonzo Bowers, who invented the first hydraulic gold dredge. **8.5" X 11", 54 ppgs, Retail Price: $9.99**

More Mining Books

Prospecting and Developing A Small Mine - Topics covered include the classification of varying ores, how to take a proper ore sample, the proper reduction of ore samples, alluvial sampling, how to understand geology as it is applied to prospecting and mining, prospecting procedures, methods of ore treatment, the application of drilling and blasting in a small mine and other topics that the small scale miner will find of benefit. **8.5" X 11", 112 ppgs, Retail Price: $11.99**

Timbering For Small Underground Mines - Topics covered include the selection of caps and posts, the treatment of mine timbers, how to install mine timbers, repairing damaged timbers, use of drift supports, headboards, squeeze sets, ore chute construction, mine cribbing, square set timbering methods, the use of steel and concrete sets and other topics that the small underground miner will find of benefit. This volume also includes twenty eight illustrations depicting the proper construction of mine timbering and support systems that greatly enhance the practical usability of the information contained in this small book. **8.5" X 11", 88 ppgs. Retail Price: $10.99**

Timbering and Mining - A classic mining publication on Hard Rock Mining by W.H. Storms. Unavailable since 1909, this rare publication provides an in depth look at American methods of underground mine timbering and mining methods. Topics include the selection and preservation of mine timbers, drifting and drift sets, driving in running ground, structural steel in mine workings, timbering drifts in gravel mines, timbering methods for driving shafts, positioning drill holes in shafts, timbering stations at shafts, drainage, mining large ore bodies by means of open cuts or by the "Glory Hole" system, stoping out ore in flat or low lying veins, use of the "Caving System", stoping in swelling ground, how to stope out large ore bodies, Square Set timbering on the Comstock and its modifications by California miners, the construction of ore chutes, stoping ore bodies by use of the "Block System", how to work dangerous ground, information on the "Delprat System" of stoping without mine timbers, construction and use of headframes and much more. This volume provides a reference into not only practical methods of mining and timbering that may be employed in narrow vein mining by small miners today, but also rare insights into how mines were being worked at the turn of the 19th Century. **8.5" X 11", 288 ppgs. Retail Price: $24.99**

A Study of Ore Deposits For The Practical Miner - Mining historian Kerby Jackson introduces us to a classic mining publication on ore deposits by J.P. Wallace. First published in 1908, it has been unavailable for over a century. Included are important insights into the properties of minerals and their identification, on the occurrence and origin of gold, on gold alloys, insights into gold bearing sulfides such as pyrites and arsenopyrites, on gold bearing vanadium, gold and silver tellurides, lead and mercury tellurides, on silver ores, platinum and iridium, mercury ores, copper ores, lead ores, zinc ores, iron ores, chromium ores, manganese ores, nickel ores, tin ores, tungsten ores and others. Also included are facts regarding rock forming minerals, their composition and occurrences, on igneous, sedimentary, metamorphic and intrusive rocks, as well as how they are geologically disturbed by dikes, flows and faults, as well as the effects of these geologic actions and why they are important to the miner. Written specifically with the common miner and prospector in mind, the book will help to unlock the earth's hidden wealth for you and is written in a simple and concise language that anyone can understand. **8.5" X 11", 366 ppgs. Retail Price: $24.99**

Mine Drainage - Unavailable since 1896, this rare publication provides an in depth look at American methods of underground mine drainage and mining pump systems. This volume provides a reference into not only practical methods of mining drainage that may be employed in narrow vein mining by small miners today, but also rare insights into how mines were being worked at the turn of the 19th Century. **8.5" X 11", 218 ppgs. Retail Price: $24.99**

Fire Assaying Gold, Silver and Lead Ores - Unavailable since 1907, this important publication was originally published by the Mining and Scientific Press and was designed to introduce miners and prospectors of gold, silver and lead to the art of fire assaying. Topics include the fire assaying of ores and products containing gold, silver and lead; the sampling and preparation of ore for an assay; care of the assay office, assay furnaces; crucibles and scorifiers; assay balances; metallic ores; scorification assays; cupelling; parting' crucible assays, the roasting of ores and more. This classic provides a time honored method of assaying put forward in a clear, concise and easy to understand language that will make it a benefit to even beginners. **8.5" X 11", 96 ppgs. Retail Price: $11.99**

Methods of Mine Timbering - Originally published in 1896, this important publication on mining engineering has not been available for nearly a century. Included are rare insights into historical methods of timbering structural support that were used in underground metal mines during the California that still have a practical application for the small scale hardrock miner of today. **8.5" X 11", 94 ppgs. Retail Price: $10.99**

The Enrichment of Copper Sulfide Ores - First published in 1913, it has been unavailable for over a century. Topics include the definition and types of ore enrichment, the oxidation of copper ores, the precipitation of metallic sulfides. Also included are the results of dozens of lab experiments pertaining to the enrichment of sulfide ores that will be of interest to the practical hard rock mine operator in his efforts to release the metallic bounty from his mine's ore. **8.5" X 11", 92 ppgs. Retail Price: $9.99**

A Study of Magmatic Sulfide Ores - Unavailable since 1914, this rare publication provides an in depth look at magmatic sulfide ores. Some of the topics included are the definition and classification of magmatic ores, descriptions of some magmatic sulfide ore deposits known at the time of publication including copper and nickel bearing pyrrohitic ore bodies, chalcopyrite-bornite deposits, pyritic deposits, magnetite-ileminite deposits, chromite deposits and magmatic iron ore deposits. Also included are details on how to recognize these types of ore deposits while prospecting for valuable hardrock minerals. **8.5" X 11", 138 ppgs. Retail Price: $11.99**

The Cyanide Process of Gold Recovery - Unavailable since 1894 and released under the name "The Cyanide Process: Its Practical Application and Economical Results", this rare publication provides an in depth look at the early use of cyanide leaching for gold recovery from hardrock mine ores. This volume provides a reference into the early development and use of cyanide leaching to recover gold. **8.5" X 11", 162 ppgs. Retail Price: $14.99**

California Gold Milling Practices - Unavailable since 1895 and released under the name "California Gold Practices", this rare publication provides an in depth look at early methods of milling used to reduce gold ores in California during the late 19th century. This volume provides a reference into the early development and use of milling equipment during the earliest years of the California Gold Rush up to the age of the Industrial Revolution. Much of the information still applies today and will be of use to small scale miners engaging in hardrock mining. **8.5" X 11", 104 ppgs. Retail Price: $10.99**

Leaching Gold and Silver Ores With The Plattner and Kiss Processes - Mining historian Kerby Jackson introduces us to a classic mining publication on the evaluation and examination of mines and prospects by C.H. Aaron. First published in 1881, it has been unavailable for over a century and sheds important light on the leaching of gold and silver ores with the Plattner and Kiss processes. **8.5" X 11", 204 ppgs. Retail Price: $15.99**

The Metallurgy of Lead and the Desilverization of Base Bullion - First published in 1896, it has been unavailable for over a century and sheds important light on the the recovery of silver from lead based ores. Some of the topics include the properties of lead and some of its compounds, lead ores such as galenite, anglesite, cerussite and others, the distribution of lead ores throughout the United States and the sampling and assaying of lead ores. Also covered is the metallurgical treatment of lead ores, as well as the desilverization of lead by the Pattinson Process and the Parkes Process. Hofman's text has long been considered one of the most important early works on the recovery of silver from lead based ores. **8.5" X 11", 452 ppgs. Retail Price: $29.99**

Ore Sampling For Small Scale Miners - First published in 1916, it has been unavailable for over a century and sheds important light on historic methods of ore sampling in hardrock mines. Topics include how to take correct ore samples and the conditions that affect sampling, such as their subdivision and uniformity. Particular detail is given to methods of hand sampling ore bodies by grab sample, pipe sample and coning, as well as sampling by mechanical methods. Also given are insights into the screening, drying and grinding processes to achieve the most consistent sample results and much more. **8.5" X 11", 124 ppgs. Retail Price: $12.99**

The Extraction of Silver, Copper and Tin from Ores - First published in 1896, it has been unavailable for over a century and sheds important light on how historic miners recovered silver, copper and tin from their mining operations. The book is split into three sections, including a discussion on the Lixiviation of Silver Ores, the mining and treatment of copper ores as practiced at Tharsis, Spain and the smelting of tin as it was practiced by metallurgists at Pulo Brani, Singapore. Also included is an overview and analysis of these historic metal recovery methods that will be of benefit to those interested in the extraction of silver, copper and tin from small mines. **8.5" X 11", 118 ppgs. Retail Price: $14.99**

The Roasting of Gold and Silver Ores - First published in 1880, it has been unavailable for over a century and sheds important light on how historic miners recovered gold and silver rom their mining operations. Topics include details on the most important silver and free milling gold ores, methods of desulphurization of ores, methods of deoxidation, the chlorination of ores, methods and details on roasting gold and silver ores, notes on furnaces and more. Also included are details on numerous methods of gold and silver recovery, including the Ottokar Hofman's Process, the Patera Process, Kiss Process, Augustin Process, Ziervogel Process and others. **8.5" X 11", 178 ppgs. Retail Price: $19.99**

The Examination of Mines and Prospects - First published in 1912, it has been unavailable for over a century and sheds important light on how to examine and evaluate hardrock mines, prospects and lode mining claims. Sections include Mining Examinations, Structural Geology, Structural Features of Ore Deposits, Primary Ores and their Distribution, Types of Primary Ore Deposits, Primary Ore Shoots, The Primary Alteration of Wall Rocks, Alterations by Surface Agencies, Residual Ores and their Distribution, Secondary Ores and Ore Shoots and Vein Outcrops. This hard to find information is a must for those who are interested in owning a mine or who already own a lode mining claim and wish to succeed at quartz mining. **8.5" X 11", 250 ppgs. Retail Price: $19.99**